# INTRODUCTION TO
# RADIOACTIVE MINERALS

Robert Lauf, Ph.D.

4880 Lower Valley Road, Atglen, Pennsylvania 19310

# Dedication

This book is dedicated to Prof. E. Wm. Heinrich (1918-1996). The thoroughness of his scholarship sets a humbling example.

Copyright © 2008 by Dr. Robert Lauf, Ph.D.
Library of Congress Control Number: 2007939782

Designed by Mark David Bowyer
Type set in Zurich BT / Zurich BT

ISBN: 978-0-7643-2912-8
Printed in China

Schiffer Books are available at special discounts for bulk purchases for sales promotions or premiums. Special editions, including personalized covers, corporate imprints, and excerpts can be created in large quantities for special needs. For more information contact the publisher:

Published by Schiffer Publishing Ltd.
4880 Lower Valley Road
Atglen, PA 19310
Phone: (610) 593-1777; Fax: (610) 593-2002
E-mail: Info@schifferbooks.com

For the largest selection of fine reference books on this and related subjects, please visit our web site at
**www.schifferbooks.com**
We are always looking for people to write books on new and related subjects. If you have an idea for a book please contact us at the above address.

This book may be purchased from the publisher.
Include $3.95 for shipping.
Please try your bookstore first.
You may write for a free catalog.

In Europe, Schiffer books are distributed by
Bushwood Books
6 Marksbury Ave.
Kew Gardens
Surrey TW9 4JF England
Phone: 44 (0) 20 8392-8585; Fax: 44 (0) 20 8392-9876
E-mail: info@bushwoodbooks.co.uk
Website: www.bushwoodbooks.co.uk
Free postage in the U.K., Europe; air mail at cost.

# Contents

# Preface

This book is intended to serve as an introduction to the world of uranium and thorium minerals for the serious student or collector; it does not claim to be truly encyclopedic in scope. (The author is presently at work on a larger volume that will cover both minerals and localities in much greater depth.) It assumes some basic knowledge of chemistry, crystallography, and geology, as well as familiarity with the commonly used terms in those fields. It also assumes the reader has a standard general reference book on mineralogy, such as one of the editions of *Dana's Mineralogy*, the *Handbook of Mineralogy*, or the *Encyclopedia of Minerals*. For this reason, the decision was made not to repeat standard quantitative data such as optical properties, density, and lattice parameters.

The general treatment begins with a brief historical discussion of the discovery of radioactive elements and the development of uranium and thorium ore deposits. Next, the geochemical conditions that produce significant deposits of primary and secondary minerals are discussed, so that when important localities are described, the reader can have a fuller understanding of their geological setting and history. Major occurrences of interest to mineral collectors are then arranged geographically.

The minerals are arranged into four broad sections, with the expectation that this treatment will be generally meaningful to most mineral collectors: 1. Primary minerals; 2. Secondary minerals; 3. Other minerals containing essential uranium or thorium; and 4. Minerals that may contain uranium or thorium as significant impurities. Emphasis is on understanding how the minerals fit into chemical groups, and for each group a few minerals are selected to illustrate their formation and general characteristics.

Because no single book can do justice to the vast literature in this field, the book concludes with a bibliography for further reading or personal research on particular subtopics and a recommended "core library" of important and reasonably accessible reference works.

# A Word About Safe Handling

Many common minerals contain toxic elements (chromium, arsenic, mercury, lead, etc.), so collectors should always exercise good hygiene when handling *any* mineral specimen: keep the materials away from food and drink and wash your hands after handling. To minimize inhalation hazards, don't create airborne dust and debris when cleaning or trimming, and thoroughly clean the work area afterwards. It is the author's belief that in general, the mere possession of a reasonable quantity of uranium or thorium minerals in a properly stored collection does not in itself present a significant additional health hazard. This belief is based on the likely dose of radiation that one would receive from this source compared to the background radiation received from all other sources. However, common sense dictates that these materials must be treated with respect. Large amounts of radioactive specimens (especially thorium minerals) should not be stored close to where people will spend a lot of time (under the bed or next to the desk, for example!) and all radioactive minerals should be carefully labeled and kept away from children. It is conceivable that a large amount of these minerals stored for long periods in an enclosed space could create a buildup of radon gas. Inexpensive radon test kits are available and any collector who is concerned about this issue should test his working or storage environment to be sure that radon is not accumulating to troublesome levels.

# Acknowledgments

The literature of radioactive minerals is broad and deep; it covers a lot of territory and many years of scientific studies as well as the careful observations of enlightened collectors. Assembling a useful and representative compendium of information from such varied sources has required a great deal of help from many friends and colleagues.

First, my supply of basic information was kept overflowing by Deborah Cole, Oak Ridge National Laboratory research librarian, who covered books, journals, and printed media in all forms.

Second, I am grateful to the following colleagues who graciously provided encouragement, technical information, reprints/preprints, and numerous helpful discussions: Thomas Armbruster, *University of Bern*; Peter Burns, *Notre Dame*; Carl Francis, *Harvard Museum*; Robert Finch, *Argonne National Laboratory*; Dermot Henry, *Museum Victoria*; Stefano Merlino, *University of Pisa*; Arvid Pasto, *Oak Ridge National Laboratory*; William "Skip" Simmons, *University of New Orleans*; Reinhard Wegner, *Universidade Federal da Paraiba*.

Third, the following mineral dealers have supplied crucial materials for my reference collection along with valuable anecdotal information on localities and history: Dudley Blauwet, *Mountain Minerals*; Richard Dale, *Dale Minerals*; Jordi Fabre, *Fabre Minerals*; Shields Flynn, *Trafford-Flynn* Minerals; the late Gilbert Gautier; Eric Greene, *Treasure Mountain Mining*; Leonard Himes, *Minerals America*; Shelley and Dan Lambert, *Lambert Minerals*; Rob Lavinsky, *Arkenstone*; Ross Lillie, *North Star Minerals*; Tom Loomis, *Dakota Matrix Minerals*; Tony Nikischer, *Excalibur Mineral Company*; Jaye Smith, *The Rocksmiths*; Sergey Vasiliev, *Systematic Mineralogy*.

Finally, I thank Sharon Cisneros, *Mineralogical Research Company*, who first kindled my special interest in radioactive minerals and who helped me build my "core collection" of specimens and scientific books. This is all her fault.

# Chapter 1
# Introduction

## • Discovery and Exploitation of Uranium and Thorium

The earliest known localities for uranium minerals are the vein deposits at Joachimsthal, Bohemia (now Jachymov, Czech Republic) and elsewhere in the Erzgebirge district. Those mines had been worked for silver and cobalt for hundreds of years, and by the early part of the 18th century pitchblende was recognized by the miners as a distinct substance, more than fifty years before the German chemist and mineralogist M. H. Klaproth showed that it contained a new chemical element. Klaproth produced "uranium" powder in 1789 by reducing pitchblende with charcoal. This product was assumed to be the element but was actually $UO_2$. About fifty years later the French chemist E. Peligot correctly identified Klaproth's material as an oxide and went on to produce elemental uranium by reducing $UCl_4$ with potassium. The French chemist Henri Becquerel discovered that uranium was radioactive in 1896, and two years later P. Curie, M. Curie, and G. Bemont isolated the element radium from pitchblende mined at Jachymov. Later work by Ernest Rutherford, Frederick Soddy, Bertram Boltwood, and others on radioactive decay pioneered modern techniques of isotopic dating. Their research also provided key evidence to refute Lord Kelvin's calculation that suggested the earth was no more than 20-30 million years old.

In 1828 the Swedish chemist J. J. Berzelius first isolated the element thorium from a silicate mineral, thorite ($ThSiO_4$), found in a pegmatite on an island in Langesund Fjord, southern Norway (Frondel 1958). M. Curie and C. G. Schmidt, working independently, discovered the radioactivity of thorium in 1889.

Modern readers naturally associate uranium and thorium with nuclear energy for power generation, weapons, and medical treatments. However, compounds of uranium have been used for over 2000 years in more traditional industries including ceramics, catalysis, and steelmaking. Colored glass dating from around 79 AD contains uranium oxide, and its use as a pigment in glass and ceramics continued well into the 20th century.

Fig.1. "Vaseline glass", shown here in natural light, gets its distinctive yellow-green color from small amounts of uranium oxide.

Fig. 2. In ultraviolet light, vaseline glass displays intense green fluorescence.

At present, the main non-nuclear use of uranium is for anti-tank weapons and other ballistic penetrators, whose performance takes advantage of the extremely high density of the material. These weapons make use of *depleted uranium*, which is essentially a very slightly radioactive waste product that remains after the fissile isotopes have been removed for nuclear fuel. Because of environmental and health concerns, depleted uranium is gradually being replaced by tungsten for ballistic applications.

Non-nuclear applications of thorium are fairly limited: its high visible emissivity when heated led to the widespread use of $ThO_2$ as a mantle in gas lanterns, and some nickel-based superalloys are strengthened by fine $ThO_2$ precipitates. Thorium is also added to tungsten cathodes and welding electrodes to decrease the electron work function, making it easier to establish and maintain a stable electron beam or arc. Over the longer term, the potential value of thorium depends on the adoption of "thermal breeder reactor" designs such as the High-Temperature Gas-Cooled Reactor (HTGR) and others. No thorium isotopes are fissile, but a thermal breeder reactor exploits the overall reaction: $^{232}Th + {}^1n \rightarrow {}^{233}U$ in which $^{232}Th$ absorbs a thermal neutron to become the unstable isotope $^{233}Th$, which decays to an isotope of protactinium and then to $^{233}U$, a fissile isotope that can be separated and fed back into a reactor to generate power. Thermal breeder reactors have been intensively researched for over 40 years, and the overall neutronics and fuel chemistry of these reactors are well understood (see, for example, Lauf et al. 1984). Prototype reactors based on thorium fuel cycles have been built in several countries, including: *AVR* in Julich, Germany; *Peach Bottom* and *Fort St. Vrain* in the US; *Dragon* project at Winfrith, UK; and *Kimini* near Kalpakkam, India.

Fig. 3. The bright orange glaze on this "Fiesta Ware" saucer contains uranium as a colorant.

# • Worldwide Economic Reserves

## Uranium

The International Atomic Energy Agency (IAEA) continuously monitors the status of the uranium industry worldwide, and issues public reports on the economic geology of uranium resources. IAEA documents of particular interest include an excellent map of the global distribution of uranium deposits (Finch et al. 1995). This map, and the accompanying guidebook, provide information on 582 uranium deposits with an emphasis on those containing at least 500 metric tons of U (tU) and an average grade of 0.03% U or greater. A biennial joint report of the OECD Nuclear Energy Agency and the IAEA, also known as the "Red Book," provides a statistical profile of the uranium industry in the areas of exploration, resource estimates, production, and reactor-related requirements.

As with all mineral resources, a particular uranium deposit may or may not be economical to develop or operate depending on the current price of uranium. For this reason, uranium reserves are generally tabulated as Reasonably Assured Resources (RAR) and Estimated Additional Resources (EAR) based on several price points: < $40/kgU; < $80/kgU; and < $130/kgU. Mines may be mothballed and later reopened as prices fluctuate, and a deposit that has been completely mined out may later yield more uranium if higher prices justify reprocessing the mine tailings to recover some of the uranium that was not extracted in the first processing. Finally, more uranium may be recovered from tailings or wastewater during environmental reclamation at a mine that has ceased production.

The following statistical "snapshot" of worldwide uranium mining activities and reserves, representing data that are believed to be current as of January 2003, is based on data from the 2003 Red Book (OECD 2004). Note that tabular data may not sum precisely because of independent rounding.

Reasonably Assured Resources among countries with major resources (tonnes U):

| Country | < $40/kgU | < $80/kgU | < $130/kgU |
|---|---|---|---|
| Australia | 689,000 | 702,000 | 735,000 |
| Kazakhstan | 280,600 | 384,600 | 530,500 |
| United States | --- | 102,000 | 345,000 |
| Canada | 297,300 | 333,800 | 333,800 |
| South Africa | 119,200 | 231,700 | 315,300 |
| Namibia | 57,300 | 139,300 | 170,500 |
| Russian Federation | 52,600 | 124,000 | 143,000 |
| Niger | 89,800 | 102,200 | 102,200 |
| Brazil | 26,200 | 86,200 | 86,200 |
| Uzbekistan | 61,500 | 61,500 | 79,600 |
| Ukraine | 15,400 | 34,600 | 64,700 |
| Mongolia | 7,900 | 46,200 | 46,200 |

Uranium deposits may be exploited by several methods, depending on the size, type, and placement of the deposit. These methods include underground mining, open-pit mining, *in situ* leaching (ISL), heap leaching or in-place leaching, and extraction as a co-product or by-product of mining for another metal.

Reasonably Assured Resources by production method (tonnes U):

| Production Method | < $40/kgU | < $80/kgU | < 130/kgU |
|---|---|---|---|
| Open-pit mining | 231,500 | 361,300 | 501,600 |
| Underground mining | 439,600 | 754,900 | 1,094,600 |
| *In situ* leaching | 358,400 | 399,400 | 450,200 |
| Heap leaching | 29,800 | 35,700 | 45,900 |
| In-place leaching | 300 | 300 | 300 |
| Co-product/by-product | 588,700 | 666,500 | 730,200 |
| Unspecified mining method | 82,100 | 240,100 | 346,400 |
| Total | 1,730,500 | 2,458,200 | 3,169,200 |

In 2002, production of uranium in North America represented about 35% of world production. Canada was the world's largest producer, at around 11,600 tU. Production in the United States was about 900 tU, most of which came from three ISL operations.

In Central and South America, Brazil was the only country producing in 2001 and 2002, and in 2002 production was about 270 tU. In Argentina the Sierra Pintada mine had been in standby status since 1999 and was expected to resume production in 2005.

In 2003 all uranium mines in Western Europe remained closed. In Germany about 200 tU were recovered from mine rehabilitation activities; small amounts were likewise recovered during mine remediation work in France and Spain.

Uranium production in Central, Eastern, and Southeastern Europe totaled around 4,200 tU, representing 12% of world production during 2002. About 500 tU were produced in the Czech Republic; a very small amount was produced in Hungary during mine reclamation. Romania produced about 90 tU. In the Russian Federation, production was about 2,800 tU in 2002 and was expected to increase significantly in 2003. Estimated production in Ukraine was about 800 tU.

Production in Africa accounted for about 17% of world production in 2002. Namibia produced around 2,300 tU, a figure that may increase by about 50% over the next several years with a new mine scheduled to open at Langer Heinrich. Niger produced about 3,100 tU. In South Africa, the uranium recovery facility at the Palabora copper mine was closed; in spite of this, production in 2002 was over 800 tU, a slight increase from 2000.

In the Middle East, Central and Southern Asia production increased to about 5,000 tU, representing about 14% of world production. Much of the increase is attributable to Kazakhstan, which produced 2,800 tU; output in Uzbekistan declined somewhat to around 1,860 tU in 2002 but was expected to increase somewhat in 2003. India and Pakistan do not report production figures, but for 2002 their output has been estimated at about 230 tU and 38 tU, respectively.

In East Asia the only uranium producing country is China, which does not report official figures; for 2002 production has been estimated at around 730 tU.

Australia, which represents about 19% of world production, suffered a decline in output during 2002 because of an accident at the Olympic Dam mine. Total production was around 6,900 tU and was expected to increase somewhat in 2003 with recent ISL operations at the Beverley mine coming on stream.

## Thorium

Thorium reserves are fairly abundant worldwide, but are not studied as intensively as those of uranium. Recent estimates of total thorium resources, which include both economic reserves and additional resources, place the figure at more than 4.5 million tonnes. Reported levels of ore reserves should be considered very conservative because of limited reporting from some countries (particularly China and the former Soviet Bloc) and because weak market demand for thorium has historically limited exploration activities (OECD 2004).

World thorium resources (economically extractable as of 1999):

| Country | Reserves (tonnes Th) |
| --- | --- |
| Australia | 300,000 |
| India | 290,000 |
| Norway | 170,000 |
| United States | 160,000 |
| Canada | 100,000 |
| South Africa | 35,000 |
| Brazil | 16,000 |
| Others | 95,000 |
| Worldwide total | 1,200,000 |

# Chapter 2
# Formation and Characteristics of Radioactive Deposits

## • Geochemistry of Uranium and Thorium

Uranium and thorium are widely distributed throughout the earth's crust. Uranium, with an average crustal abundance of 2.4 ppm, is more plentiful than such familiar elements as silver, gold, arsenic, mercury, molybdenum, antimony, and tin. Thorium, at 12 ppm, is roughly as plentiful as lead and boron.

In primary rocks and magmas, uranium is normally present in the quadrivalent state ($U^{4+}$). During early crystallization of the magma, the low concentration of U prevents it from crystallizing as distinct mineral phases. At the same time, the large size and charge of the $U^{4+}$ ion effectively keep it from becoming incorporated in the usual rock-forming silicate minerals as they crystallize. Thus, uranium tends to become concentrated in the hydrothermal solutions and pegmatites that remain as the solidifying magma becomes consolidated. Small amounts of uranium may be taken into solid solution in various accessory minerals, particularly zircon, monazite, and thorite. The bulk of the uranium will ultimately crystallize as uraninite, $UO_2$, although in some hydrothermal vein deposits the primary uranium mineral is coffinite, $U(SiO_4)_{1-x}(OH)_{4x}$. Within the pegmatites, uranium may be present in solid solution in thorite, thorianite, monazite, zircon, brannerite, and various niobate-tantalate minerals. In these deposits, uranium is normally associated with thorium, zirconium, and rare-earth elements (REE). By contrast, in hydrothermal deposits of magmatic origin, uranium has become separated from these elements and is more typically associated with cobalt, nickel, bismuth, and arsenic. In these settings uraninite is again the most abundant primary uranium compound, but it typically forms as fine-grained, massive *pitchblende*.

The geochemical behavior of uranium has important similarities and differences compared to that of thorium, cerium, and zirconium, and it is instructive to examine these in detail (Frondel 1958). The similarities arise from their general chemical behavior, particularly their strong affinity for oxygen. The dissimilarities relate to the small size of the $Zr^{4+}$ ion compared to $U^{4+}$, $Th^{4+}$, and $Ce^{4+}$ and, more importantly, to the fact that only uranium has a higher valence state, $U^{6+}$. In their quadrivalent states, all of these elements show the same general behavior in magmas and tend to occur together, either as separate accessory minerals in the same pegmatite or as substitutional solid solution species in the same mineral.

In contrast to thorium, cerium, and zirconium, only uranium has a stable hexavalent state, usually forming the uranyl ion, $(UO_2)^{2+}$; importantly, this is the stable oxidation state under normal atmospheric conditions. This ion is a linear $[O\text{-}U\text{-}O]^{2+}$ complex that behaves as a distinct unit in crystal structures. Its large "dumbbell" shape prevents it from substituting for other divalent ions, giving rise to a whole collection of unique secondary minerals in which the uranyl ion is a crucial part of the structure. Furthermore, the uranyl ion tends to be fairly soluble, so once primary uraninite is weathered or otherwise oxidized, the uranyl species can migrate over large distances before precipitating as carbonates, phosphates, sulfates, silicates, or vanadates. Because $Th^{4+}$ has no higher valence state that would correspond to $(UO_2)^{2+}$, the primary mineral thorianite, $ThO_2$, is quite stable in the environment and there are no corresponding secondary thorium minerals.

When primary uraninite crystallizes, even in an oxygen-poor environment, it always contains some oxygen in excess of the stoichiometric ratio, that is, the formula would be expressed as $UO_{2+x}$, where x ranges from about 0.15 to perhaps as much as 0.6. Once uraninite is exposed to the air, or to oxygenated groundwater, further oxidation proceeds steadily. Oxidation kinetics is made faster by the breakdown of the uraninite crystal structure from internal radiation damage (*metamictization*) as well as by the process of *auto-oxidation*. The concept of auto-oxidation is based on the fact that the products of radioactive decay, primarily lead, do not have as high an affinity for oxygen as uranium does; thus, as more lead is produced within the uraninite, the uranium ions react to what appears to be a more oxidizing condition by moving from $U^{4+}$ to $U^{6+}$. If sulfur is available, the radiogenic lead typically forms galena, often seen as microscopic inclusions within the uraninite.

The mobility of uranium in the environment is highly dependent on its oxidation state, with $U^{6+}$ being considerably more mobile than $U^{4+}$. Based on extensive thermodynamic calculations, Langmuir (1978) has shown that $U^{4+}$ can only reach significant concentrations in ground water as the fluoride complex. This complex is only stable in reduced, acidic waters (pH below about 3) containing sufficient fluoride. In the natural environment, ground waters satisfying all of these conditions are very rare and consequently $U^{4+}$ may be generally regarded as substantially immobile.

When uraninite becomes oxidized and water is present, a number of soluble uranyl complexes can be formed depending on factors such as the pH and the other anions that are present in the water. Some soluble uranyl species are listed in the following table. It must be kept in mind that in an aqueous solution there will usually be several ionic species present at different concentrations (defined by the equilibrium constant of the particular reactions), so the table presents the dominant species in a particular pH range:

| Water acidity | Acid (pH < 4) | Near neutral (pH = 6) | Neutral to alkaline (pH > 7) |
|---|---|---|---|
| Dominant ionic species | $(UO_2)^{2+}$ | $(UO_2)OH^+$ | $(UO_2)_3(OH)_5^+$ |
| | | $(UO_2)SiO(OH)_3^+$ | |
| | | $(UO_2)(HPO_4)_2^{2-}$ | $(UO_2)(HPO_4)_2^{2-}$ |
| | | $(UO_2)CO_3$ | $(UO_2)(CO_3)_2^{2-}$ |

The solubilities of uranyl minerals follow a fairly regular sequence in order of decreasing solubilities: carbonates > sulfates > phosphates and arsenates > silicates > vanadates. It is not surprising, then, that the most commonly found uranyl minerals are the vanadates carnotite and tyuyamunite, followed by autunite-group phosphates and arsenates, and silicates, particularly uranophane. The large number of potential combinations of the uranyl ion with other metals and each of the major anions creates a fascinating diversity of known secondary minerals, and new ones continue to be described today.

Fig. 4. Cubic crystals of uraninite about 1 cm across, altering to orange curite. An orange alteration rim is also visible on the massive black uraninite at lower left. Specimen is from Shinkolobwe, Congo. *RJL1700*

Fig. 5. Rear of the specimen shown in Figure 4, showing successive layers of alteration products on the uraninite; the outermost layer is a patch of golden-yellow uranophane needles. *RJL1700*

# • Classification of Radioactive Mineral Deposits

The inherent complexity of uranium deposits inevitably frustrates human attempts to create simple, consistent classification systems. Many schemes have been proposed and each has its own merits and shortcomings. Purely descriptive terms based on the local geology are less likely to provoke controversy; as McMillan (1978) points out, "...genetic ideas have a tendency to change while the observed wallrock association does not!" On the other hand, the main purpose of studying ore deposits is to understand their genesis so that exploration for similar deposits might be focused on geological settings that have the greatest chance of being productive. Thus it is most common to classify deposits based on both the geologic setting and the probable mechanisms of formation, with the understanding that in many cases the geologic evidence might be subject to more than one equally plausible genetic interpretation.

For the mineral collector, as opposed to the exploration geologist, a general understanding of the geology will aid the interpretation of a mineral specimen: What are the associated minerals? What is the matrix? Is the specimen typical of the locality? Does the specimen nicely illustrate some unique features of that locality? How much geologic "history" is preserved and displayed by the sample at hand? The following simplified discussion of uranium deposits is based largely on the classification used by Heinrich (1958).

*1. Syngenetic Deposits in Igneous Rocks:* In granitic and intermediate rocks, most of the radioactivity lies in the accessory minerals zircon, titanite, apatite, allanite, xenotime, and monazite. Minor radioactive accessory minerals include uranothorite, thorianite, euxenite, thorite, pyrochlore, chevkinite, fluorite, bastnaesite, and davidite. The radioactivity usually varies significantly across a single pluton, with concentration of uranium and thorium often seen at the outer margins of the pluton and/or in the last material to crystallize, as expected based on the large size and charge of $U^{4+}$ and $Th^{4+}$.

*2. Radioactive Pegmatite Deposits:* Uranium and thorium are fairly common minor constituents in pegmatites. As with the plutonic deposits described previously, pegmatites are rarely of economic value as uranium ore but are the main source for many of the primary uranium and thorium minerals and the U- and Th-containing complex oxides sought by collectors. Pegmatites typically occur in "swarms" or well-defined groups associated with a larger pluton that represents the source of the original magma. Many radioactive species are found in pegmatites; some are primary minerals formed directly from magmatic or hydrothermal processes, whereas others are secondary minerals formed *in situ* through the action of rain or ground water. The primary radioactive minerals often found in pegmatites include uraninite and to a much lesser extent thorianite. The complex mixed oxides (pyrochlore and samarskite groups, the aeschynite series, davidite, chevkinite, and others) are often found, and in fact pegmatites are the main source of many of these minerals. Phosphates such as monazite and, more rarely, xenotime, as well as silicates such as thorite, allanite, gadolinite, and zircon are fairly widespread primary pegmatite minerals. Well-known localities for primary minerals include the region around Bancroft, ONT; Charlebois Lake and Foster Lake areas, SASK; the Ruggles pegmatite, Grafton, NH; and near Idaho Springs, CO. In Northern Karelia, Russia, there are at least five separate pegmatites that contain uraninite (Heinrich 1958).

Most secondary uranium minerals found in pegmatites can be attributed to the action of air and water on the primary minerals. The most commonly seen species include autunite, torbernite, and uranophane, along with a mixture of alteration products called *gummite*; other hydrous uranyl oxides, carbonates, arsenates, and silicates are sometimes found in pegmatites. Noteworthy examples of this type of deposit are the localities in New England; near Spruce Pine, NC; the region between Sabugal and Guarda, Portugal; and Salamanca, Spain.

*3. Carbonatites:* Carbonatites are magmatic carbonate-silicate rocks that contain mainly calcite or dolomite (and sometimes other carbonate minerals) with lesser amounts of silicates, oxides, and phosphates, and which are thought to have been deposited from a carbonate-rich fluid. They are almost always associated with nepheline syenite ring complexes, which can range up to several km across. Carbonatites often form the lower part of the necks of some explosive-type volcanoes, but can also form other structures such as dikes, cone sheets, and breccia zones. Often these structures are surrounded by a contact metasomatic aureole in which the original country rocks have been altered to syenites by the action of alkalic solutions, a process called *fenitization*. A flow structure can sometimes be seen in the parallel orientation of biotite, apatite, rutile, or amphibole crystals.

Carbonatites have relatively high concentrations of phosphorus, chlorine, fluorine, titanium, zirconium, niobium, cerium-group REE, sulfur, barium, and strontium, along with appreciable thorium and (to a lesser degree) uranium. They are mineralogically complex; typical radioactive species include pyrochlore, monazite, thorite, bastnaesite, thorogummite, xenotime, and betafite. Notable localities include: Alno Island, Sweden; the Fen area, near Telemark, Norway; Tororo, Uganda; Oka, QUE, Canada; thorium deposits in Gunnison Co., CO; and at the Sulfide Queen body in San Bernardino Co., CA.

**4. Pyrometasomatic Deposits:** Pyrometasomatic deposits of uranium and thorium minerals are not especially common. Within this broad category are many subtypes, including the following:

*Deposits of thorianite-uraninite in marble, as masses of lime-rich amphibolite ("metapyroxenite") near intrusive contacts.* In the marble a skarn mineral assemblage is typically seen, and the recrystallized calcite is often salmon-pink. The chief radioactive species is uranoan thorianite or thorian uraninite; minor amounts of thorite, monazite, allanite, melanocerite, and other REE species may be found. In the lime-rich amphibolites the usual radioactive species is uraninite; pyrochlore and uranothorite are sometimes seen. Examples are found in the Bancroft, ONT region, Canada; Easton, PA; and the Ft. Dauphin area in Madagascar.

*Contact allanite deposits* occur in contact zones between intrusives and various sedimentary deposits. In some of these occurrences allanite and other REE minerals are abundant but only weakly radioactive, whereas in others the allanite is responsible for considerable radioactivity. Deposits of this type include the Whalen mine in west-central Alaska; Grand Calumet Is., QUE; the Kyshtymsk district, Ural Mts., Russia; and at Bastnaes and at the Ostanmossa mine, Sweden. At the Mary Kathleen deposit, QSL, Australia the allanite itself is not radioactive, but disseminated grains of uraninite are present and in the oxidized zone uranophane, uranophane-ß, and *gummite* are found.

*Metasomatic monazite deposits* found in northern Lemhi Co., ID consist of monazite-bearing carbonate layers in a country rock of mica schists and gneisses. The carbonate lenses consist of calcite, stained in places by hematite and containing monazite along with actinolite, barite, siderite, pyrite, magnetite, ilmenorutile, garnet, zoisite, and apatite.

The honey-colored monazite forms fine to coarse, generally euhedral crystals and irregular masses in concentrated zones that can reach several meters thick and 100 m long. The overall mineral assemblage suggests a metasomatic origin (Heinrich 1958) but the deposits also show some similarities to some carbonatite vein deposits in Montana.

*Oxidized contact deposits* contain various secondary minerals such as autunite, metatorbernite, and metazeunerite. In some localities (e.g., near Austin, NV; near Deer Lodge, MT; and at Brooks Mt., Seward Peninsula, AK) the original contacting bodies and the resulting "primary" contact zone mineralogy are fairly well understood. A notable exception is the important deposit at Mt. Spokane, WA; there, the primary phases are not well understood and details of the genesis of this deposit remain incompletely known.

**5. Hypothermal Deposits:** In hypothermal deposits, formed at 300-500°C, uranium and thorium tend to be closely associated in the minerals davidite, brannerite, and thorian uraninite. Common associate minerals include ilmenite, hematite, pyrite, magnetite, and molybdenite. Other minerals that are sometimes seen are huebnerite, cobaltite, cassiterite, and arsenopyrite. Some examples of hypothermal deposits include the following:

*Davidite veins* are exemplified by the deposit at Radium Hill, SA, Australia. The deposit is postulated to have developed in several stages: 1. Replacement of micaceous shear rock by quartz-biotite-hematite-ilmenite mineralization; 2. Intrusion of REE-rich pegmatites containing allanite and xenotime; 3. Movement along the shears, causing brecciation and biotization; 4. Intrusion of amphibolites along faults; and, 5. Introduction of quartz stringers containing davidite. The davidite also developed intergrowths with the existing ilmenite-hematite mineralizations. Davidite vein deposits also occur at several sites in the Mavuzi district, Mozambique, where davidite crystals exceeding a foot in diameter were found.

*Brannerite deposits*, presumed to be of hypothermal origin, occur at the California mine, Chaffee Co., CO; at the West Walker River deposit, near Coleville, CA; and in the molybdenite deposit at Chateau-Lambert, Vosges, France.

**6. Mesothermal Deposits:** Mesothermal deposits, which formed at intermediate temperatures and depths (200-300°C), include several very productive mining districts such as Great Bear Lake, NWT, Canada; Shinkolobwe, Congo; and Goldfields, SASK, Canada, where large tonnages of high-grade ore have been recovered. There are three main types of radioactive mesothermal deposits: Ni-Co-Ag type; pitchblende-pyrite type; and monazite or thorite veins. The Ni-Co-Ag type deposits are further sub-divided by some authors into numerous subtypes based on details of the mineral assemblage.

In general, mesothermal vein deposits usually contain a gangue of carbonate minerals and/or coarse quartz. Hematite is very common and often abundant enough to color the gangue minerals and impregnate the wall rock. Pyrite and chlorite are common; barite and fluorite, along with arsenopyrite and other sulfosalts, are rarer. Many of these deposits contain significant amounts of gold, silver, or platinum-group metals (PGM). The usual base metals are bismuth, copper, lead, cobalt, and nickel; zinc and molybdenum are sometimes found as well. The predominant copper mineral is usually chalcopyrite; bismuth is typically present as either the native element or bismuthinite. The cobalt and nickel minerals vary greatly from one deposit to another and may include niccolite, cobaltite, gersdorffite, glaucodot, skutterudite, smaltite, chloanthite, safflorite, rammelsbergite, maucherite, polydymite, vaesite, cattierite, and siegenite. At Shinkolobwe, selenium is abundant in the Co-Ni minerals, particularly as selenian digenite, whereas at Goldfields, selenides of copper and lead are found.

The normal radioactive mineral in mesothermal deposits is uraninite (var. pitchblende) that is usually low in thorium but may contain REE. From measurements of the $UO_2$ unit cell parameters, the estimated $U^{4+}/U^{6+}$ ratio indicates formation at intermediate temperatures. In most cases the uraninite is botryoidal or spherulitic; a notable exception is at Shinkolobwe, where it is granular to euhedral in small cubic crystals.

Some examples of Ni-Co-Ag uranium deposits are: the Radium vein at the Caribou deposit near Nederland, CO; the Black Hawk silver district, NM; the Coeur d'Alene district, ID; Great Bear Lake area, NWT, Canada; Cornwall, England; the Erzgebirge district; and the Shinkolobwe mine, Congo.

Examples of pitchblende-pyrite type deposits include: the Central City district, Gilpin Co., CO; Theano Point, Montreal River district, ONT, Canada; the Goldfields region, SASK, Canada; Placer de Guadalupe, CHIH, Mexico; Margnac, France; the Darwin-Katherine area, NT, Australia; Witwatersrand, South Africa; and Serra de Jacobina, BAH, Brazil.

*Thorium veins* are the least common type of radioactive mesothermal deposit. In the Lemhi Pass district of Idaho and Montana, the metamorphic country rock (argillite, quartzite, and mica schist) is cut by fissures that are filled by copper veins. These veins contain bornite, chalcopyrite, chalcocite, pyrite, quartz, hematite, gold, silver minerals, and thorite. Associated with these are quartz-hematite veins containing barite, goethite, and thorite. Thorogummite is sometimes present in the iron oxides and small amounts of monazite and allanite may be present in some of the veins.

A *monazite deposit* at Steenkampskraal, South Africa contains dark veins of corroded and sericitized feldspar with a fine-grained mixture of quartz, pyrite, hematite, monazite, apatite, and leucoxene. In some places the vein material is as much as 75% monazite and the ore contains as much as 5% $ThO_2$. The mineralization is interpreted as having occurred in three general stages (Heinrich 1958): 1. Feldspar, monazite, apatite, magnetite, and possibly ilmenite; 2. Quartz, pyrite, and possibly chalcopyrite; and 3. Sericite.

The *radioactive zirconium deposits* of Pocos de Caldas, Brazil contain an unusual assemblage of baddeleyite/zircon mixtures ("caldasite") containing up to several percent uranium, as well as magnetite ore containing veins of bastnaesite and thorogummite. The presence of baddeleyite, along with zeolite pockets, suggests that the mineralization took place under conditions that fall within the mesothermal range.

*7. Epithermal Deposits:* Uraniferous vein deposits formed by low-temperature (100-200°C) hydrothermal processes may be quartzose, fluoritic, or both, and are relatively sulfide-poor. The usual uranium species is botryoidal to fine-grained pitchblende. Individual sites may be completely silicic, completely fluoritic, or virtually any intermediate composition. In a few cases, for example in the Thomas Range, UT, a single fluorite pipe grades downward into a chalcedony replacement mass. The gangue minerals may include clays, alunite, iron and manganese oxides, and small amounts of calcite or dolomite. Pyrite and marcasite are often present, along with small amounts of galena, chalcopyrite, and sphalerite, but overall the veins contain fairly small amounts of metallic sulfides.

Epithermal thorite veins contain thorite and thorogummite and relatively little uranium. The gangue minerals include feldspar, smoky quartz, siderite, fluorite, barite, and hematite.

*Siliceous pitchblende-sulfide veins* occur at the Gray Eagle mine and other lead-silver deposits in the Boulder Batholith, MT. The gangue material is primarily quartz and the ore is mainly pitchblende (replacing pyrite). In several mines in the Boulder district secondary minerals are found in the oxidized parts of the deposits; these include uranophane, uranophane-ß, meta-autunite, metatorbernite, metazeunerite, and phosphuranylite. Important deposits in Portugal and France are also examples of this type.

*Massive fluorite deposits* in the Thomas Range, UT do not have commercially significant uranium concentrations, but at several of the sites minor amounts of secondary minerals such as carnotite are found in the upper parts.

*Fluorite-sulfide veins* in the Jamestown, CO district were long mined for fluorite. Localized concentrations of both uraninite (pitchblende) and uranothorite occur in purple fluorite breccia pipes. Secondary deposits of torbernite are also found in the area. Veins of fluorite, pitchblende, pyrite, and ilsemannite are found at the Miracle Mine and the Kergon deposit in Kern Co., CA; autunite, torbernite, and other uranyl minerals are locally abundant. At the Rexspar deposit, BC, Canada, the fine-grained uraninite contains thorium and REE and is associated with fluorite, celestite, mica, and pyrite. The hydrothermal concentration of radioactive species, REE, fluorine, and strontium is interpreted to be of volcogenic origin (McMillan 1978).

*Fluorite-quartz-sulfide veins* in the Marysvale, UT district contain chalcedony or coarse quartz along with fluorite, adularia, and hematite. Pitchblende, pyrite, marcasite, and jordisite are present, as well as mercury minerals of the metacinnabar-tiemannite series. In the oxidized zone, numerous secondary minerals are found, including autunite, torbernite, metatorbernite, phosphuranylite, uranophane, uranophane-ß, schroekingerite, johannite, uranopilite, zippeite, tyuyamunite, and rauvite. Fluorite vein deposits in the Wölsendorff, Germany region are also examples of this type.

*Thorite veins and breccia pipes* in the Wet Mountains, Custer Co., CO, contain quartz, feldspar, barite, and iron oxides, with smaller amounts of fluorite, siderite, and various sulfides. There are finely disseminated veinlets and pods of thorite and thorogummite, and xenotime can be locally abundant. Many REE are present but uranium concentrations are quite low.

**8. Deposits in Sedimentary Rocks:** Sedimentary-hosted uranium deposits include three major types: 1. sandstone; 2, quartz-pebble conglomerate; 3. unconformity. These deposits are rarely the source of good specimens for the collector, but are of great importance because they form the vast bulk of economic ore reserves at present. Their economic value lies partly in their large size and partly in the low cost of exploiting them. In some cases the uranium is recovered by a process called *in-situ leaching*, in which wells are drilled to the porous host layer. Solutions are pumped in some of the wells to create soluble uranyl species, which are then pumped out of adjacent wells.

*Sandstone-type deposits* include those often referred to as "Colorado Plateau type". Examples occur throughout the Colorado Plateau itself as well as in Wyoming, South Dakota, Texas, and California; similar occurrences have been discovered in Argentina and Russia. Within this category, however, individual deposits show numerous variations in their mineralogy and probable origin.

Peneconcordant deposits lie roughly parallel to the bedding of the host sandstone. In these structures the main ore minerals, uraninite and coffinite, tend to be closely associated with organic material, which might be carbonized or coalified wood, plant debris, or humic material. The soluble uranyl species were presumably captured by adsorption onto the organic matter and then chemically reduced by the carbonaceous material and other reducing species such as $H_2S$, which could have been biogenically derived.

Roll front deposits, sometimes called geochemical cell deposits, usually have a crescent-shaped cross section that can extend laterally for some distance. In this type of deposit, concentrations of uraninite and coffinite lie at or near oxidation-reduction interfaces where oxygenated, uranium-rich groundwater came into contact with a reducing environment containing carbonaceous material and pyrite. The changing redox conditions (and pH) cause the precipitation of reduced uranium species and sometimes vanadium and molybdenum minerals in the interface zone (De Voto 1978). These deposits were probably formed below about 50°C and it is likely that communities of microorganisms played a role in reducing the soluble $U^{6+}$ to form the insoluble uranium minerals (Suzuki and Banfield 1999).

The mineralogy of sandstone-type deposits is complex and depends, among other things, on the presence or absence of vanadium. As Heinrich (1958) points out, vanadium provides a convenient internal measure of the geochemical conditions because it has such a wide variety of minerals each of which is only stable over a narrow range of oxidation potential. In nonvanadiferous ores, numerous secondary uranium minerals may be found, including oxides, sulfates, carbonates, members of the autunite group, and various uranyl silicates. In vanadiferous ores the uranium minerals include carnotite, tyuyamunite, uranophane, autunite, zippeite, schroeckingerite, andersonite, becquerelite, and schoepite. If the ore is incompletely oxidized, other vanadium minerals may include paramontroseite, corvusite, melanovanadite, duttonite, simplotite, sherwoodite, and doloresite. If the ore is fully oxidized, the vanadium minerals tend to include navajoite, hewettite, pascoite, rossite, metarossite, hummerite, steigerite, fervanite, volborthite, calciovolborthite, and pintadoite.

*Quartz-pebble conglomerate deposits* represent very ancient alluvial formations, or *paleoplacers*. At the time these deposits were formed, roughly 2-3 B years ago, there was much less free oxygen in the atmosphere, which allowed the $U^{4+}$ species (uraninite or brannerite) to survive long enough for the placer to become buried under additional sedimentary layers, which then protected them from oxidation as the atmosphere evolved to its present state. The typical mineral assemblage is quartz, gold, pyrite, uraninite, brannerite, zircon, chromite, monazite, leucoxene, and platinum-group minerals (PGM) such as osmium-iridium alloys, isoferroplatinum, and sperrylite. In some cases the gold and PGM values determine the economics of mining. Examples of conglomerate deposits are: Witwatersrand, South Africa; Elliot Lake, Canada; Jacobina, Brazil; and Tarkwa, Ghana.

*Unconformity-type deposits* were first recognized as a distinctive class in the early 1970s and are represented by major uranium mining districts in northern Saskatchewan, Canada and in Northern Territory, Australia. Together, these districts represent about 16% of the world's uranium reserves (Nash et al. 1981). The two deposits have important similarities and differences, and the question of how they were formed continues to be debated. Briefly, both deposits lie at or near an unconformity where Lower Proterozoic metasediments are overlain by Middle Proterozoic quartz sandstone. The mineralization seems to be locally controlled by faulting, fracturing, or collapse breccias.

Mineralogy of the northern Saskatchewan deposit is fairly complex. For example, the mineralization at Rabbit Lake is postulated to have occurred in three stages (Hoeve and Sibbald 1978). Stage 1: veins of pitchblende and coffinite with adularia, chlorite, quartz, hematite, calcite, quartz, and sulfides; chloritization and hematization of the wall rock indicates oxidizing ore fluids. Stage 2: formation of veins of euhedral quartz with little uranium under somewhat reducing conditions. Stage 3: impregnation of sooty pitchblende and coffinite along fractures, also under reducing conditions.

Mineralogy of the Northern Territory deposit is somewhat simpler and wall rock alterations are more uniform, suggesting that mineralization took place under uniformly reducing conditions (Edwards and Atkinson 1986). The primary ore mineral is uraninite, with lesser amounts of coffinite, brannerite, and *thucolite* (a generic term for radioactive mineral hydrocarbon assemblages). The gangue mineral is predominantly chlorite, with hematite plentiful at some sites. At the Ranger 1 mine, economically significant secondary minerals (mainly saleeite, along with sklodowskite, gummite, and metatorbernite) occur in a thick oxidation zone.

**9. Placer Deposits:** The relative ease with which uraninite is oxidized virtually precludes even highly crystalline material from surviving long enough to be transported intact to a modern placer deposit. Thorium minerals are stable with respect to oxidation, but they are frequently metamict and this tends to reduce their resistance to weathering and alteration. As a result, most radioactive placer deposits of commercial interest contain mixtures of various minerals such as monazite, zircon, thorite, euxenite, samarskite, and xenotime. Individual rounded grains of these minerals make up the bulk of the "heavy sand" concentrates, whose overall composition reflects both the source rock and the intensity of weathering and transport.

A placer deposit of some interest to collectors lies in the interior of Sri Lanka. In the Peak district, small, unabraded black cubes of thorianite are recovered as a byproduct of placer gem mining. This occurrence has been known since the early 1900s and yielded over nine tons of thorianite (Heinrich 1956).

**10. Deposits Formed by Weathering and Groundwater:** This category includes many deposits that are of great interest to mineral collectors because it is here that the interesting geochemistry of uranium is best displayed in the formation of colorful secondary minerals. These minerals may develop directly upon or near the primary deposit as a coating, replacement, or pseudomorph. Alternatively, soluble uranyl ions may travel great distance before precipitating when the water chemistry changes.

*Veins oxidized in place* typically begin as a vein or other hydrothermal deposit containing pitchblende, often with pyrite or other metallic sulfides. Oxidation of the sulfides introduces sulfuric acid, $H_2SO_4$, into meteoric or ground water, which readily attacks the uraninite. The uranium is preferentially leached as soluble $(UO_2)^{2+}$. Radiogenic lead and radium tend to form very insoluble sulfates and are therefore not mobilized, but remain as minute radioactive fracture fillings in the corroded uraninite. The most common secondary mineral species that form tend to be uranyl phosphates, sulfates, and silicates. Often, $Ca^{2+}$ and $Cu^{2+}$ are present as well, so autunite, torbernite, and uranophane are particularly widespread examples. Other commonly-seen species include uranocircite, saleeite, parsonsite, kasolite, zippeite, uranopilite, and schroekingerite. Vanadates such as carnotite are fairly rare unless the primary ore minerals contain vanadium. If the primary vein material contains significant arsenic, then arsenate minerals such as zeunerite and uranospinite may form.

The secondary uranyl oxides such as schoepite, masuyite, and becquerelite tend to form directly on the uraninite as alteration layers or pseudomorphs. Where there is a lot of ground water flow, these minerals tend not to survive in place for very long periods but instead are ultimately leached away.

The thickness or depth of the oxidized zone can vary considerably; for example, at Shinkolobwe, unoxidized ore begins at a depth of about 50 m, whereas in the Gunnar deposit, Goldfields, SASK, secondary minerals (particularly uranophane) are found at a depth of about 300 m. In some cases the lower limit of the oxidation zone is fairly well defined by the water table, but there are many exceptions to this generality. When oxidation extends for a great distance below the present water table it may be attributed to several possibilities: 1. a seasonally fluctuating water table; 2. large variations in permeability created by localized fractures; 3. a water table that was previously much lower but has since been raised by climatic changes (Heinrich 1958).

Important examples of oxidized vein deposits include: Spokane, WA; Marysvale, UT; Goldfields, SASK, Canada; Shinkolobwe, Congo; and many localities in Portugal and France.

*Deposits formed by groundwater redistribution* result when mobile $(UO_2)^{2+}$ complexes encounter an environment favorable for precipitation or adsorption. In some cases, the original radioactive source rock is not known with certainty. Uraniferous coals, lignites, and black shales are found in large areas of the western U.S. as well as in Europe and Madagascar. These deposits rarely contain recognizable uranium minerals. As noted earlier, groundwater redistribution also plays an important role in many Colorado Plateau type deposits.

# • Provinces and Epochs

There is ample evidence that certain petrographic provinces have their own characteristic radioactivity, just as others might have notable enrichments in other chemical elements. For example, acid igneous rocks from the western margin of the Canadian Shield (e.g., Great Slave Lake and Great Bear Lake areas) are about five times as radioactive as acid rocks from the interior parts of the shield in Ontario and Quebec. Various authors have proposed the concept of a "uranium metallogenic province" as a broad but indefinitely bounded area in which uranium-rich rocks are relatively abundant; the uranium deposits within the province might be of more than one type and age (Klepper and Wyant 1956). This concept suggests that the non-uniform distribution of radioactive species arose early in the development of the earth's crust and persists as large-scale features to the present time.

Similarly, the ages of many uranium deposits seem to be concentrated in three major groups: Laramide and Tertiary; Hercynian; and Precambrian, although there were likely several separate epochs of uranium mineralization during the Precambrian.

The major uranium provinces are summarized by Heinrich (1956) as:

1. The Colorado Plateau and the surrounding arc of vein deposits that run from the Black Hills at the northeast end, down through the Colorado Front Range, westward through New Mexico and Arizona, then northward through Marysville, Utah and into Idaho.

2. A narrow belt along the southern and western margins of the Canadian Shield, encompassing the Bancroft area, Blind River, Theano Point, west-central Manitoba, Goldfields, and Great Bear Lake.

3. A strip along the eastern side of Brazil, particularly an area of Rio Grande do Norte and Paraiba; and an area to the southeast in Bahia, Minas Gerais, Espirito Santo, and Sao Paulo.

4. A broad, generally east-west strip across Central and Western Europe, encompassing the Erzgebrige, Wölsendorf, and Wittichen in Germany; the Massif Central in France; Cornwall, England; and the deposits of Portugal.

5. A wide, discontinuous belt through Southern Africa that includes occurrences in the Congo; Zimbabwe; Witwatersrand in South Africa; Mozambique; and Madagascar.

6. A large area in Australia that encompasses Darwin, Northern Territory; the Mt. Isa-Clooncurry region in Queensland; Broken Hill, New South Wales; and Radium Hill and Mt. Painter, South Australia.

7. The Ferghana – Kara Tau region of the Former Soviet Union.

# Chapter 3
# Important Radioactive Mineral Localities

## • North America

### United States

There are about a thousand documented radioactive mineral deposits in the United States, and many have been thoroughly studied (Heinrich 1958). Several districts may be considered classic localities based on their historical importance, quality of the specimen material, and familiarity to the collecting community.

In New England, "swarms" of radioactive pegmatites have been known since the late 1800s. Two distinctive types of specimens are especially noteworthy: At the Swamp mine, Topsham, ME uraninite forms sharp, complex crystals, typically up to a centimeter across. At the Ruggles pegmatite, Grafton Center, NH uraninite forms dendritic aggregates in white feldspar; colorful "gummite" crusts are formed where these rocks are exposed to surface weathering.

Fig. 6. A sharp single uraninite crystal about 1 cm tall with a bit of feldspar from the Swamp mine, Maine. *RJL2281*

Fig. 7. A large (10 X 14 cm) slab of dendritic uraninite with yellow/orange "gummite" from the classic locality at the Ruggles mine, Grafton Center, New Hampshire. *RJL3139*

The radioactive pegmatite at Spruce Pine, NC continues to yield excellent specimens of nicely crystallized torbernite as well as other colorful secondary minerals such as uranophane, meta-autunite, and fourmarierite.

In the western U.S., the Colorado Plateau region is home to commercially important uranium ore deposits and is also a rich source of uranium minerals for the collector. Species first described from the Colorado Plateau region (and the type locale, TL, from which each was first described) are: abernathyite (Fumerole No. 2 mine, Emery Co., UT); andersonite, bayleyite, and swartzite (Hillside mine, Yavapai Co., AZ); blatonite and oswaldpeetersite (Jomac mine, San Juan Co., UT); boltwoodite (Pick's Delta mine, Emery Co., UT); haynesite and larisaite (Repete mine, San Juan Co., UT); holfertite (Starvation Canyon, Thomas Range, UT); metatyuyamunite (Jo Dandy mine, Montrose Co., CO); rabbittite (Lucky Strike No. 2 mine, Emery Co., UT); umohoite (Freedom No. 2 mine, Piute Co., UT); and weeksite (Autunite No. 8 claim, Juab Co., UT).

Fig. 8. Bright green flakes of torbernite on pegmatite from Spruce Pine, North Carolina. *RJL2862*

Commercially significant deposits of autunite and uranophane were discovered in the Mt. Spokane area, WA in the 1950s. Rich mineralized veins extend along a contact between granite and argillites, and in places autunite is found in the granite as much as 6 m from the contact. The Daybreak mine, in particular is noteworthy for the quality and extremely large size of its autunite crystals. In the best specimens, dark green tabular crystals form beautiful fanlike aggregates or "books" several cm across, and are arguably the world's finest examples of the species.

## Canada

Uranium mining in Canada began with the discovery in 1930 of the Port Radium deposit on the eastern shore of Great Bear Lake, NWT. The deposit is primarily mesothermal veins of the Ag-Co-Ni type with Bi and Cu, containing veins and stringers of pitchblende. The Eldorado mine was worked from 1933 to 1940

Fig. 9. Rich green autunite from the Daybreak mine, Spokane, Washington. *RJL659*

Fig. 10. The autunite specimen in Fig. 9, shown in ultraviolet light. *RJL659*

and reopened in 1942 to supply uranium to British and American defense programs. The nearby Contact Lake mine was a small producer of native silver and pitchblende, and many similar deposits have been reported from the district (Heinrich 1958).

The Blind River-Elliot Lake district, ONT, was discovered in 1949 and produced uranium for more than forty years. The deposit is a quartz pebble conglomerate and although the ore was fairly low grade, the favorable location and low processing costs made the deposit economical to exploit (Heinrich 1958). The last production facility at Elliot Lake closed in 1996.

At present, all active uranium production is from unconformity-related deposits in northern Saskatchewan. Within that district, the McArthur River mine is the world's largest uranium mine. Canada is presently the world's top uranium producer; its annual output, around 11,000 tU, represents about 30% of the world total.

---

For collectors, the unique calcite-fluorite-apatite vein-dikes in the Haliburton-Bancroft area, ONT, are of particular interest. Within the intrusives and the metamorphosed Grenville limestones are found several dozen radioactive species, of which fine crystals of betafite from the Silver Crater mine are perhaps best known to collectors. The Madawaska mine (formerly Faraday mine) has produced world-class uranophane and uranophane-ß specimens. Excellent simple or modified cubic uraninite crystals are found at the Cardiff mine. Other species noted from the district include: euxenite, fergusonite, thorianite, thorite, uranpyrochlore, and zircon var. *cyrtolite*, a radioactive variety characterized by curved faces and pale tan color (Kennedy 1979).

Fig. 11. Gray crystal of uraninite in calcite with fluorite and apatite from the Cardiff Uranium mine, Wilberforce, Ontario, Canada. *RJL579*

Fig. 12. Intergrown cluster of betafite crystals from the Silver Crater mine, Bancroft, Ontario, Canada. *RJL337*

## Mexico

The Sierra Peña Blanca district, which lies 50 km north of Chihuahua City, is an example of uranium deposits related to volcanic rocks. Three isolated deposits have been studied extensively and each has its own characteristic genesis: Nopal I is a hydrothermal type; Las Margaritas is a mixed supergene-exhalative type; and Puerto 3 is a supergene type (George-Aniel et al. 1991). Many secondary uranium species are found in the Peña Blanca deposits, particularly uranophane, uranophane-ß, and weeksite. Las Margaritas is the type locale for margaritasite, the Cs analogue of carnotite (Wenrich et al. 1982).

Fig. 14. Yellow earthy margaritasite filling mm-sized vugs in matrix from the type locale, Margaritas open pit, Sierra Peña Blanca, Chihuahua, Mexico. *RJL418*

Fig. 13. Pale yellow matted aggregates of acicular uranophane-ß from the Faraday mine, Bancroft, Ontario, Canada. *RJL218*

The Moctezuma mine (locally called "La Bambolla"), near Moctezuma, Sonora, is a gold-tellurium deposit that was mined for gold in the 1930s and 40s. A large number of new minerals have been described from this mine and the nearby San Miguel mine, including three uranyl tellurites: cliffordite, schmitterite, and moctezumite. Elsewhere in Sonora, excellent specimens of torbernite are found at La Luz.

Fig. 15. Tiny patches of yellow cliffordite on matrix, associated with dark olive-green mackayite (an iron tellurite) from Moctezuma, Sonora, Mexico. *RJL338*

Fig. 16. Pearly green tabular crystals of torbernite from La Luz, Sonora, Mexico. *RJL3076*

# • South America

## Brazil

Uranium prospecting, which began in 1952, led to the discovery of significant deposits at Pocos de Caldas in Minas Gerais and at Jacobina in Bahia. Further exploration in the 1970s led to the discovery and evaluation of eight deposits, of which Osamu Utsumi and Lagoa Real were developed into mining projects. Although Brazil holds significant uranium resources, annual mine output is fairly low, partly because the economics of some large phosphorite deposits will depend on the markets for the phosphate minerals from which uranium will be extracted as a byproduct.

Several noteworthy localities in Brazil have produced fine specimens for collectors:

The Corrego do Urucum pegmatite, Galileia, Minas Gerais is the type locale for coutinhoite; phosphuranylite, saleeite, uraninite, uranocircite, meta-uranocircite, weeksite, and wölsendorfite are also found there.

The uranium deposits at Teofilo Otoni, MG, have produced superb specimens of haiweeite, along with autunite, parsonsite, and uranocircite. Elsewhere in Minas Gerais, well-crystallized autunite is found at Malacacheta and at Sao Jose da Safira.

The granitic pegmatite at Perus, SP has yielded numerous radioactive species, including: autunite, meta-autunite, bassettite, chernikovite, haiweeite, phosphuranylite, phurcalite, torbernite, metatorbernite, uranophane, uranophane-ß, and weeksite. Fine crystals of novačekite are found at the Brumado mine, BAH, along with sklodowskite, thorutite, zeunerite, and metazeunerite.

**Fig. 17. Bright chartreuse meta-autunite crystals to about 4 mm, scattered on dark mica, associated with white feldspar from São Jose da Safira, Minas Gerais, Brazil.** *RJL3110*

Fig. 18. Bright yellow acicular crystals of haiweeite forming 2-3 mm spherical tufts on calcite from Teofilo Otoni, Minas Gerais, Brazil. *RJL2719*

# • Europe

## England

There are no commercially important uranium deposits in the British Isles; all uranium minerals were extracted as a by-product of mining for other metals. Tin and copper accounted for the bulk of commercial production, along with significant amounts of lead, silver, tungsten and arsenic (Embrey and Symes, 1987). Uranium and many other industrial metals were recovered in lesser amounts. For the collector, several Cornish mines have produced fine specimens of torbernite and bassettite among others.

A large number of Cornish mines yielded uranium minerals (Embrey and Symes 1987) while they were operating and their waste dumps continue to yield specimens to this day. Wheal Edward has produced pitchblende along with zeunerite and other secondary minerals. Commercial quantities of pitchblende were mined at Wheal Trenwith in the early 1900s. The Tincroft mine, which dates from the early 1700s and was closed in 1913, has yielded torbernite. Wheal Basset is one of the "classic" localities of Cornwall; it is the type locale for both bassetite and vochtenite and has also produced specimens of torbernite and uranospathite. Torbernite has been found at Wheal Butler, which was an important copper producer until it was abandoned in the market slump of the late 1870s. The Tolcarne mine, which was worked for tin in the early 1600s, produced some good specimens of torbernite. Small amounts of torbernite and autunite had been known from the area of South Terras; around 1880 a rich vein was found and this mine became a leading producer of uranium for the next 30-40 years. Nearby, St. Austell Consols, a tin mine, also produced commercial quantities of uranium ore. Wheal Trewavas is the type locale for tristramite. The Old Gunnislake mine, which was started in the 1700s, produced the best of the Cornish torbernite specimens.

At present, there is no active metal mining in Cornwall, but the hundreds of abandoned mine sites are still of great interest. The British government is identifying and protecting remaining sites that are considered to be of special importance, and in these places uncontrolled collecting or other disturbance is now limited by law. At the same time, many dumps are being destroyed by land reclamation as well as by reprocessing the waste to recover the metal values that remain. Nonetheless, the ore and waste dumps remain a valuable source of materials for hobby collectors and for ongoing scientific study. It is interesting to note that vochtenite was first described from Wheal Basset (Zwaan et al. 1989) *seventy years* after underground workings had ended there.

Fig. 19. A very old specimen of torbernite from the Gunnislake mine, Cornwall, England. This specimen was in the collection of Archduke Stephan of Austria ca. 1850-60. It was later obtained by a European museum through the Rumpff collection in 1889. *RJL1929*

Fig. 20. Another view of the preceding specimen, showing stacked formation of the torbernite crystals. *RJL 1929*

Fig. 21. Pearly micaceous plate of greenish yellow bassetite, associated with yellow phosphuranylite from the type locale, Wheal Basset, Cornwall, England. *RJL480*

## Portugal

Uranium development began in 1912 with the discovery of the Urgeiriça U-Ra deposit. Radium was mined until 1944 and uranium production began in 1951. Urgeiriça, along with about 100 nearby deposits, formed an important mining district running in an arc from Coimbra to Viseu to Guarda. Another cluster of deposits lies about 50 km south, near Portalegre. All of the Portugese deposits are granitic and were mined by either underground or open-pit methods. Several mines were listed as producing in the mid-1990s (Finch et al. 1995); however, more recently all mines appear to be dormant and reclamation was expected to have started in 2003.

For the collector, some classic material has come from Portugal, including fine autunites and metatorbernites. Sabugal, Portugal is the type locale for sabugalite, another member of the autunite group.

Fig. 22. Green tabular metatorbernite on matrix from Forte Velho, near Guarda, Portugal. *RJL247*

Fig. 23. Stacks of thin tabular autunite crystals forming a solid crust about 5 X 6 cm, from Sabugal, Portugal. *RJL660*

## Spain

Radioactive veins associated with granite like those in Portugal extend southeastward into Spain. The earliest uranium discoveries were in Salamanca in the late 1950s, followed by additional deposits further south in Badajoz. Although the supply of high-quality specimen material from Spain is generally less than that of Portugal, locales in Badajoz have produced specimens of uraninite (var. pitchblende) and colorful secondary minerals including autunite, sabugalite, saleeite, torbernite, metatorbernite, and uranospathite.

## France

Uranium exploration began in 1946 and continued at a high level until 1987, after which uranium production began to decline. About thirty significant deposits were developed, most of which were epithermal veins associated with granite. The last operating mines in France were Lodeve (closed in 1997) and Le Bernardan (closed in 2001). Because of the large number of localities and the long history of uranium mining there, France has been a rich source of superb specimens of many radioactive species. Uranium minerals first described from France (with their TLs) are: autunite and meta-autunite (Autun, Saone-et-Loire); compregnacite (Margnac, Haute-Vienne); deliensite (Mas D'Alary, Lodeve); deloryite (Cap Garonne, Var); fontanite and rabejacite (Rabejac deposit, Lodeve); and rameauite (Margnac, Haute-Vienne).

Fig.24. Tiny pale yellow platy crystals of sabugalite on dark granitic matrix from Olivia de la Frontera, Badajoz, Spain. *RJL2648*

Fig. 25. Bright greenish yellow autunite crystals scattered on matrix from the type locale at Autun, France. *RJL3051*

Fig. 26. Superb group of tabular crystals and "books" of torbernite to 1 cm+, forming a solid plate about 8 cm long, from Margabal, Aveyron, France. *RJL2417*

Fig. 27. Orange granules of earthy marecottite from the type locale, La Creusaz, near Les Marecottes, Switzerland. *RJL3058*

## Switzerland

At present, there are no commercially active uranium mines in Switzerland. However, the deposit at La Creusaz, near Les Marecottes village in the Western Alps, is of some interest to collectors as the type locale for marecottite; it is scientifically important as an example of the active formation of secondary minerals as a result of the interactions between primary minerals and acid mine drainage. The La Creusaz prospect was discovered in 1973 and explored by drilling and by digging shallow trenches and tunnels. The primary mineralization contained uraninite and pyrite, with minor amounts of Bi, Pb, Se, and other species. The type specimen of marecottite was described from a specimen that had formed on stockpiled ore in an exploration tunnel where it had been left since 1981 and subject to acid mine drainage waters (Brugger et al. 2003). Other recently formed secondary sulfates found in this material include members of the zippeite group, jachymovite, and several potentially new minerals.

## Germany

In the years prior to reunification in 1991, East Germany was a major uranium producer, with total cumulative output estimated at over 200,000 tU. All mines have now ceased operation, although reclamation work continues to yield small quantities of uranium. For example, decommissioning of the Königstein mine in Saxony in 1999 yielded around 30 tU.

Germany has been a rich source of fine specimens for collectors. The Streuberg mine, near Bergen an der Trieb, Vogtland, Saxony is the type locale for bergenite, meta-uranocircite I and II, phurcalite, and uranocircite; autunite, billietite, meta-autunite, phosphuranylite, and renardite are also found there. The Krunkelbach Valley deposit, near Menzenschwand in the Schwartzwald Mts., is the type locale for joliotite, uranosilite, and uranotungstite; several dozen uranium minerals have been reported there. Also in the Schwartzwald Mts., Wittichen is the type locale for orthowalpurgite.

Fig. 28. Bright yellow mass of platy bergenite in a quartz-lined vug from Mechelgrun, near Bergen, Vogtland, Germany. *RJL2248*

Fig. 29. A fine example of uranocircite from Streuberg, Saxony, Germany. *RJL3112*

## Czech Republic

The Erzgebirge ("Ore Mountains") forms a natural boundary between Saxony in Germany and Bohemia in the Czech Republic. Mining in this region began with the discovery of silver at Freiberg in 1170 and has continued to the present. In the 1700s mining was directed mainly to cobalt for ceramic colorants. In the 1800s, important products were cobalt and bismuth, with silver as a valuable by-product. Uranium ore was first mined at Jachymov strictly as a source of radium. Under Soviet control in the period following World War II, peak production of uranium from the Erzgebirge represented perhaps 3-4% of worldwide production (Heinrich 1958).

Ore deposits in the district are generally classified as veins associated with granite, and include: pyrometasomatic iron-copper deposits; tin-tungsten veins; pyritic lead veins; silver-cobalt-bismuth-uranium mineral veins; and veins of iron and manganese oxides. In these veins, all of the uraninite is pitchblende. Weathering of these ores has produced a fascinating array of secondary minerals (Ondrus et al. 2003) and further research might well yield additional new species from this prolific region.

Jachymov is the type locale for twenty-six minerals, among which are the following fourteen radioactive species, listed in their order of discovery from 1727 to the present: uraninite, torbernite, johannite, zippeite, voglite, uranopilite, schroeckingerite, uranophane-ß, meta-uranopilite, nickel-zippeite, albrechtschraufite, jachymovite, čejkaite, and pseudojohannite.

Other radioactive species reported from Jachymov include: andersonite, arsenuranospathite, autunite, becquerelite, boltwoodite, coffinite, compreignacite, cuprosklodowskite, curienite, dewindtite, kahlerite, kamotoite-(Y), kasolite, liebigite, magnesium-zippeite, marecottite, masuyite, meta-autunite, metakirchheimerite, metalodevite, metanovacekite, metaschoepite, metatorbernite, metatyuyamunite, meta-uranocircite, meta-uranospinite, metazellerite, metazeunerite, novacekite, parsonsite, phosphuranylite, rabbittite, rabejacite, richetite, schoepite, sklodowskite, soddyite, sodium-zippeite, torbernite, troegerite, tyuyamunite, urancalcarite, uranocircite, uranophane, uranospathite, uranosphaerite, uranospinite, vandendriesscheite, vochtenite, walpurgite, weeksite, widenmannite, wölsendorffite, yingjiangite, zellerite, zeunerite, and znucalite.

Fig. 30. Bright green, elongated tabular crystals of johannite from the type locale at Jachymov, Czech Republic. *RJL379*

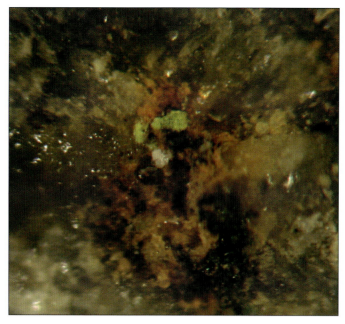

Fig. 31. Minute (~0.5 mm) patch of green pseudojohannite on ore from the type locale, Jachymov, Czech Republic. *RJL3066*

## • Russia and Central Asia

### Russia

Intensive exploration, beginning in 1944, identified over 100 uranium deposits within fourteen districts of the Russian Federation. With cumulative production (through 2002) of about 120,000 tU the Russian Federation ranks as the fifth largest uranium producer in the world. Several other republics of the former Soviet Union host significant uranium ore deposits as well as interesting mineral assemblages. Many radioactive species were first described from Kazakhstan, Kyrgyzstan, Tadjikistan, and Uzbekistan, as well as from Russia; however, locality information was often missing or inaccurate during the Cold War era. Recent academic work in Russia (Pekov 1998; Pekov 2007) has provided welcome details to fill the gaps in our knowledge of all these localities.

Radioactive species first described from Russia (and their TLs) are: aeschynite-(Ce), chevkinite-(Ce), and samarskite-(Y) (Ilmeny Mts., S. Urals); bauranoite, calciouranoite, and metacalciouranoite (Oktyabrskoye Mo-U deposit, near Krasnokamensk, E. Transbaikal); diversilite-(Ce) (Yukspor Mt., Khibiny massif, Kola Peninsula); iriginite and moluranite (Aleksandrovskii Golets Mo-U deposit, Udokan Range, N. Transbaikal); karnasurtite-(Ce) and umbozerite (Karnasurt Mt., Lovozero massif, Kola Peninsula); lermontovite (Beshtau uranium deposit, near Pyatigorsk, N. Caucasus); plumbobetafite (Burpala massif, Mama River basin, Siberia); pseudo-autunite (Vuoriyarvi massif, N. Karelia); vanuranylite (Ust-Yuk V-Se-U deposit, Tuva, Siberia); yttrobetafite-(Y) and yttropyrochlore-(Y) (Alakurtti granite pegmatite, N. Karelia).

Fig. 32. Dark needles of umbozerite in alkaline igneous rock from Mt. Panqurayv, Lovozero massif, Kola Peninsula, Russia. *RJL1495*

## Kazakhstan

Uranium has been mined in Kazakhstan since 1957, when open-pit mining began at the Kurdai deposit. For the next twenty years, about fifteen separate deposits, primarily vein-type, were mined by either open-pit or underground methods and yielded about 5,000 tU in total. Mining of extensive sandstone-type deposits by ISL began in 1978 and current production is estimated to be about 3,000 tU/year (OECD 2004). Radioactive species first described from Kazakhstan (and their TLs) are: chistyakovaite, sodium uranospinite, and uramarsite (Bota-Burum uranium deposit, Chu-Ili Mts., SW Balkhash region); mourite and sedovite (Kyzylsai Mo-U deposit, Chu-Ili Mts., SW Balkhash region); saryarkite-(Y) (Akkuduk rare-metal deposit, near Mointy railway station); and sodium boltwoodite (unnamed uranium occurrence within the Kyzylsai ore field, SW Balkhash region).

## Kyrgyzstan

Two small underground mines at Mailisai and Mailisu, Osh region produced uranium from a limestone-bitumen deposit; both are now depleted. Radioactive species first described from Kyrgyzstan (and their TLs) are: tyuyamunite (Tyuya-Muyun Cu-V-U deposit, Ferghana Valley, Alai Range); and uramphite (Tura-Kavak uranium-coal deposit).

## Tadjikistan

A small vein-type deposit at Taboshary was mined by underground methods and is now depleted. Radioactive species first described from Tadjikistan (and their TLs) are: arapovite (Darai-Pioz Glacier, Alai Range); bismutopyrochlore (Mika pegmatite vein, Tau Mt., Eastern Pamirs); calcioursilite and magnioursilite (Oktyabrskoye uranium deposit, Kyzyltyube-Sai, near Khodzhent); chernikovite (Karakat uranium deposit, Karamazar Mts.); przhevalskite (Dzerkamar uranium deposit, near Andrasman, Karamazar Mts.); and sodium autunite (Kuruk uranium deposit, near Khodzhent).

## Uzbekistan

Several small volcanic vein-type deposits in the Ferghana valley and the Kazamazar uranium district were mined beginning around 1946 and are now depleted. All of the country's important uranium resources are sandstone or black shale type deposits in the central Kyzylkum area. Several dozen sandstone-type deposits have been identified, many of which would be amenable to in-situ leaching. Seven black shale deposits have been found, which could be mined by open pit and extracted by heap leaching methods. Open pit and underground mining began at Uchkuduk in 1958 and ISL extraction was applied on a commercial scale beginning in 1965. Currently, all uranium production in Uzbekistan is by the ISL method (OECD 2006). Radioactive species first described from Uzbekistan (and their TLs) are: arsenuranylite (Cherkasar uranium deposit, near Pap); and vyacheslavite (Dzantuar and Rudnoye uranium deposits, Auminzatau Mts., Central Kyzylkum region).

Fig. 33. Large crude crystal of davidite-(La) from Bektau-Ata, Kazakhstan. *RJL2575*

Fig. 34. Minute orange aggregates of arsenuranylite from the type locale, Cherkasar uranium deposit, near Pap, Uzbekistan. Associated minerals are schoepite (yellow) and zeunerite (green). *RJL1674*

# • East Asia and Australia

## Sri Lanka

There are no commercial uranium resources in Sri Lanka, but several radioactive minerals are obtained from heavy sand and placer deposits. The island has produced hundreds of tons of monazite from beach deposits and several tons of thorianite from river gravels. The monazite in Sri Lanka has unusually high Th content (about 10% $ThO_2$), compared to monazites from India (8%) and Brazil (5%). Small thorianite deposits are fairly widespread in the interior highlands; this thorianite typically contains 10-30% $U_3O_8$ (Nininger 1954). Isolated thorianite crystals are recovered from river gravels as a by-product of gem mining, as relatively unabraded black cubes. In the Peak district, *nine tons* of this material was reportedly recovered during the period from 1904 to 1910 (Heinrich 1958).

in SA. Olympic Dam has been operating since 1988; it is the world's largest known uranium deposit and is Australia's largest underground mine. Uranium is produced along with gold as a co-product of copper mining. Beverley, the second-largest uranium resource in Australia, is a sandstone type deposit; there uranium is extracted by in-situ leaching. The Ranger open pit mine, operating since 1981, is an unconformity-related deposit in which the main ore mineral is *pitchblende*, but it is best known to collectors for superb saleeite crystals.

The Mt. Painter deposit, Northern Flinders Ranges, SA, is of great historical interest and although it was never a large commercial success it has produced a great variety of secondary uranium minerals. Documented species include billietite, boltwoodite, françoisite-(Nd), schoepite, metaschoepite, soddyite, metatorbernite, uranophane-ß, and weeksite. It is the type locality for spriggite (Brugger, Ansermet, and Pring 2003). Nearby Radium Hill is the type locality for davidite-(La); several secondary minerals are also found there, including carnotite and uranospinite.

Fig. 35. Intergrown group of 3-mm cubic thorianite crystals from the Balangoda district, Sri Lanka. *RJL331*

## Australia

Uranium was first produced in Australia in 1906 at Radium Hill, SA. Exploration picked up around 1944 and led to the development of the Rum Jungle, South Alligator River Valley, Radium Hill, and Mary Kathleen deposits. Further exploration in the late 1960s led to the discovery of major deposits at Lake Frome and the Ngalia basin (Langford 1978). In the 1970s major discoveries included Ranger, Jabiluka, and Nabarlek in NT, Yeelirri in WA, and Olympic Dam

Fig. 36. Bright yellow tablular saleeite crystals thickly scattered on matrix from the Ranger mine, Northern Territory, Australia. *RJL2863*

Fig. 37. Dull yellow coating of carnotite on massive black davidite-(La) from Radium Hill, South Australia, Australia. *RJL1330*

The South Alligator Valley uranium field, NT, comprises over twenty identified deposits, many of which were actively mined for uraninite during the 1950s and '60s. In some mines, important gold values were also encountered. More recent interest in uranium during the 1980s led to the discovery of significant gold, platinum, and palladium mineralization at Coronation Hill. The rare species threadgoldite is widespread in the district (Henry et al. 2005). Other secondary minerals noted there include autunite, meta-autunite, becquerelite, curite, dewindtite, dumontite, kasolite, masuyite, phosphuranylite, rutherfordine, saleeite, soddyite, torbernite, metatorbernite, uranophane, uranopilite, and vandendriesscheite (Threadgold 1960).

The Lake Boga Granite, VICT, is the type locale for ulrichite; several other secondary species are found there, including torbernite, saleeite, and metanatroautunite ("sodium autunite").

Fig. 38. Bright green divergent spray of ulrichite about 2 mm across, associated with smoky quartz in granite from the type locale at Lake Boga, Victoria, Australia. *RJL2890*

## • Africa

### Congo

The sheer quantity and quality of its specimens has made Congo the premier locality for students of uranium mineralogy. It should be noted that two neighboring countries in West Africa go by the name of "Congo." The country of interest here is the former Belgian Congo, which was later named Zaire, and later still re-named Democratic Republic of the Congo (DRC). To further complicate matters, the name of the copper-rich Katanga province in the southern part of the country was changed to Shaba province but now is once again generally called Katanga. (Both names are derived from tribal words for copper.) Collectors may encounter various combinations of these names on labels, particularly with older specimens.

The rich copper deposits of Katanga have been known and exploited by natives since ancient times. Within this area are several significant uranium occurrences, each with its own distinctive geology and suite of secondary minerals.

The first documented find of uranium was at the Luiswishi copper mine in 1913. However, the amount of uranium ore was not economically important. The major deposits at Shinkolobwe were discovered in 1915 and were mined for radium from about 1921 through 1936. Mining was resumed in 1944 and continued until 1959. Long after the mine was closed, scientists have continued to describe new minerals from archived ore samples (Deliens et al. 1981).

Uranium was found at the Kalongwe copper mine associated with chalcocite and chalcopyrite orebodies in 1930. Although uranium has not been actively mined at this locale, it has produced specimens of numerous secondary uranium minerals.

The deposit at Swambo was discovered in 1956, somewhat by accident, as it was not part of a copper mine or prospect. Swambo is the type locality for swamboite and has also yielded fine soddyite crystals.

The Kamoto East mine is noteworthy because of a unique combination of uranium, copper, and rare earth elements that yielded several new U-REE species (Deliens et al. 1990).

The Musonoi uranium-copper-cobalt deposit contains an interesting suite of secondary uranyl selenates in the surface alteration zone. The selenium is apparently derived from weathering of copper sulfide ores that contain some Se and Te in substitution for sulfur.

The uraniferous pegmatite at Kobokobo, Kivu Province, which was mined for beryl and columbite, produced at least a half-dozen new aluminum uranyl phosphates and several containing thorium as well.

Well over fifty uranium minerals were first described from localities in Congo. From Kalongwe: vandenbrandeite. From Kamoto East: astrocyanite-(Ce), françoisite-(Nd), kamotoite-(Y), and shabaite-(Nd). From Kobokobo: althupite, eylettersite, kamitugaite, kivuite, metavanmeerssheite, moreauite, mundite, phuralumite, ranunculite, threadgoldite, triangulite, upalite, and vanmeerssheite. From Musonoi: demesmaekerite, derricksite, guilleminite, marthozite, and sengierite. From Shinkolobwe: becquerelite, bijvoetite-(Y), billietite, cuprosklodowskite, curite, dewindtite, dumontite, fourmarierite, ianthinite, kasolite, lepersonnite-(Gd), masuyite, metaschoepite, metastudtite, oursinite, paraschoepite, parsonsite, piretite, richetite, roubaltite, saleeite, sayrite, schoepite, sharpite, sklodowskite, soddyite, studtite, urancalcarite, vandendriesscheite, and wyartite. From Swambo: swamboite.

Fig. 39. Brilliant green needles of cuprosklodowskite forming a geode in cavernous matrix from Kolwezi, Congo. *RJL2617*

Fig. 40. Detail of the cuprosklodowskite nodule showing associated minerals: malachite (very dark green); rutherfordine (pale cream yellow); and schoepite (bright golden yellow). *RJL2617*

Fig. 41. Bright green scaly marthozite crystals on massive selenian digenite from the type locale, the Musonoi mine, Kolwezi, Congo. *RJL1145*

## Gabon

The uranium-vanadium deposits of Mounana, in the province of Haut-Ogoue, were discovered in 1956 by geologists of the French Atomic Energy Commission. In all there are six U deposits in the Franceville basin, of which four (Boyindzi, Mounana, Oklo, and Okelobondo) were economically important, with total U resources estimated at 40,000 tons of U (Gauthier-Lafaye and Weber 1989). Open pit mining began in 1961 and annual production peaked at around 1250 tons of uranium in the late 1970s. The last areas mined were an underground working at Okelobondo and an open pit at Mikouloungou, which closed in March, 1999. Total uranium production was about 26,000 tons. The Franceville basin is especially noteworthy for two reasons. First, the oxidized zone at Mounana has produced an extraordinary variety of uranyl vanadates and many new vanadium minerals. Second, very rich orebodies at Oklo, Okelobondo, and Bangombe formed natural nuclear reactors about 2 billion years ago, a phenomenon that has not been observed anywhere else in the world.

The Mounana Mine is well known to collectors as a classic source of secondary uranium minerals. It is the type locality for mounanaite, $PbFe_2^{+3}(VO_4)_2(OH)_2$, and superb specimens of yellow francevillite on velvety brown mounanaite are among the unique and sought-after mineral assemblages found only at this mine. The Mounana deposit forms a fault wedge in coarse sandstones of Middle Cambrian age. The ore body was about 150 m long, up to 40 m wide, and 150 m deep. The mine was worked as an open pit to a depth of 70 m, and then by underground methods. The ore deposit formed two distinct zones (Cesbron and Bariand 1975):

1. The deep or "black ore" zone where U and V are in their lower oxidation states and form one mineral assemblage. Within the black ore zone there is a uranium-rich area where the main ore minerals are uraninite and some coffinite. Together with galena, sphalerite, iron and copper sulfides, and asphaltic material these minerals form the sandstone cement. Accessory minerals in this zone included zircon, tourmaline, rutile, and barite. Also within the black ore zone was a vanadium-rich pillar containing a number of rare vanadium oxides and hydroxides, including karelianite, montroseite, roscoelite, nolanite, duttonite, lenoblite, and bariandite.

2. The oxidized zone, about 40 m thick, where the mineral suite was predominantly aluminum, lead, and barium uranyl vanadates. The oxidized zone contained an extraordinary suite of uranyl vanadates, typically filling cracks up to several cm thick, in the friable sandstone. From this zone seven new minerals were described in the period from 1957 to 1971. The following species have been recovered from this deposit: carnotite, francevillite, curienite, vanuralite, metavanuralite, chervetite, mottramite, mounanaite, schubnelite, uranocircite, torbernite, kasolite, uranopilite, and johannite. The oxide zone was completely mined out in the mid-1970s.

## Namibia

Namibia is currently the fourth largest uranium producer in the world. Production capacity at the Rössing mine is about 4,000 tU/year, with actual production running around 2,500 tU/year. Several other significant deposits exist, the most economically favorable being the low-grade surficial duricrust (hard-pan) deposit at Langer Heinrich, which would be mined as an open pit with a projected capacity around 1,000 tU/year (OECD 2004).

The Rössing deposit is located in the Namib desert about 40 miles northeast of Swakopmund. The presence of radioactive minerals was known as early as 1928, based on simple autoradiography tests of rocks from the area. Detailed exploration programs were conducted off and on through 1973, and mine development began in 1974.

The main orebody at Rössing is a uraniferous alaskite rock, whose texture ranges from aplitic to granitic to pegmatitic, with the pegmatitic texture generally predominating. All of the primary minerals and most of the secondary minerals occur within the alaskite. The primary minerals (uraninite with minor betafite) represent about 60% of the uranium present. The remaining 40% is contained in the secondary minerals, which include *gummite*, uranophane, uranophane-ß, torbernite, metatorbernite, carnotite, metahaiweeite, and thorogummite (Berning et al. 1976).

Fig. 42. A choice example of brilliant orange francevillite on dark brown mounanaite from the type locale for both minerals, the Mounana mine, Haute-Ogoue, Gabon. *RJL1353*

Fig. 43. Yellow acicular crystals of boltwoodite forming divergent sprays on black calcite from Rössing, Namibia. *RJL1301*

The Rössing deposit, along with nearby prospects, is an important source of fine materials for the collector. The locality is best known for colorful boltwoodite specimens, with radiating acicular crystals ranging from opaque straw yellow to translucent orange forming spherical aggregates on dark calcite. The area has also yielded a small number of brilliant yellow weeksite specimens. Although still "micro" crystals, they are spectacular examples of the species.

Fig. 44. Bright yellow weeksite from an ore prospect near the uranium mine at Rössing, Namibia. *RJL2704*

## Madagascar

Madagascar is notable for zoned pegmatite deposits containing complex oxides and other radioactive minerals including fergusonite, samarskite, euxenite, betafite, allanite, chevkinite, uraninite, thorite, and monazite. Antsirabe, Madagascar is the type locale for betafite, which is abundant at many places on the island and occasionally forms large crystals and crystal groups. Crystals up to 7 cm (and one crystal group weighing 104 kg) have been reported (Heinrich 1958).

Fig. 45. A thumbnail-sized euxenite-(Y) crystal group from Betafo, Madagascar. The crystals are covered with a thin layer of tan surface alteration products; euxenite-(Y) is very dark when fresh. *RJL3077*

# Chapter 4
# The Minerals

## • Primary Minerals

### Oxides

The thorianite-uraninite system forms a complete series of solid solutions with the cubic crystal structure from $ThO_2$ to $UO_2$, based on studies of synthetic materials. In nature, however, compositions with Th:U ratios close to 1:1 are relatively rare (the "uranothorianites" of Madagascar are close to a 1:1 composition).

Uraninite occurs in several distinct geological settings: 1. In pegmatites, typically as well-formed crystals or dendritic growths associated with monazite, columbite-tantalite, and zircon; 2. In high temperature hydrothermal veins, typically as massive pitchblende associated with cassiterite, arsenopyrite, and Co-Ni-Bi-As minerals; 3. In moderate-temperature hydrothermal veins with galena and other sulfides; and 4. In sandstone deposits with V, Cu, and other metal ores.

In pegmatites, the uraninite crystals usually contain appreciable levels of thorium (up to ~10%), rare earths (up to ~15%), and other impurities, because the size and valence of these ions prevent them from entering into the main rock-forming silicates as they crystallize. Pitchblende, on the other hand, is usually fairly pure because it has crystallized from hydrothermal fluids that don't contain these elements. Both varieties of uraninite may contain a significant amount of radiogenic lead (up to ~20%) depending on the geologic age of the deposit (Frondel 1958). A few noteworthy locales are of interest to the collector: Lustrous black crystals are found at Topsham, ME and Spruce Pine, NC. Dendritic uraninite in pegmatite, usually associated with colorful alteration products (*gummite*) is well known from the Ruggles pegmatite, Grafton, NH. Pitchblende is found in vein deposits at Jachymov, Czech Republic (TL) and in tin-bearing veins in Cornwall, England. Pitchblende is also a frequent component of sandstone or Colorado Plateau type deposits.

**Fig. 46. Simple cubic uraninite crystal in calcite with dark grains of fluorite and a yellow-green apatite crystal from the Cardiff Uranium mine, Wilberforce, Ontario, Canada.** *RJL578*

Fig. 47. Dendritic uraninite in pegmatite with orange *gummite* from the Ruggles mine, Grafton Center, New Hampshire. *RJL3133*

Fig. 48. Massive black uraninite var. pitchblende from Moab, Utah. *RJL267*

Fig. 49. Typical black simple cubic thorianite single crystal about 9 mm across, from Fort-Dauphin, Madagascar. *RJL632*

Thorianite occurs chiefly in pegmatites and granites, or in sediments and placers derived from these deposits. It typically contains 5-10% $UO_2$ and up to 8% REE oxides, as well as up to 5% radiogenic lead. Because $Th^{4+}$ has no higher valence (corresponding to $U^{6+}$) thorianite is chemically stable in the environment, although the crystals are usually metamict to some degree and can undergo further alteration as any substitutional $U^{4+}$ is oxidized to $U^{6+}$. Thorianite usually forms simple black cubic crystals and penetration twins. In the Balangoda district, Sri Lanka (TL), sharp unaltered crystals are recovered from stream gravels as a byproduct of gem mining.

## Silicates

The primary uranium and thorium silicates include coffinite, huttonite, thorite, and thorogummite. Of these, coffinite, thorite, and thorogummite are tetragonal members of the zircon group. Huttonite is the monoclinic dimorph of thorite and is structurally related to the phosphate minerals monazite and cheralite.

Fig. 50. Rough tan crystal of thorite var. uranothorite from Bancroft, Ontario, Canada. The crude, rounded lozenge shape is typical of samples from this locality. *RJL157*

Thorite, considered as the theoretically pure end member, would be ThSiO$_4$, by analogy to zircon. However, thorite nearly always contains significant uranium, and the accepted formula is (Th,U)SiO$_4$. Thorite occasionally forms large, usually brown, crystals that are rarely as sharp as those of zircon because of alteration and metamictization. Major substitutional impurities are U (up to ~25%), REE (up to ~8%), PO$_4$ (up to ~13%), and radiogenic Pb. Good crystals are found at several locales in the Bancroft, ONT district, Canada and at Arendel, Norway. The variety *orangeite*, found at Aroplany, Madagascar, was named for its distinctive orange-yellow color.

Fig. 51. Group of elongated gray-brown prismatic crystals of thorite var. uranothorite in pegamatite from Pacoima Canyon, San Gabriel Mts., Los Angeles Co., California. The resemblance to zircon can be seen in the pyramidal termination at top of the front-most crystal. *RJL566*

Fig. 52. Crude, brick-red crystals of thorite var. orangeite in pegmatite, associated with purplish zircon, from Aroplany, Tulear, Madagascar. *RJL3126*

Coffinite and thorogummite have long been regarded as isostructural with thorite but with some $(SiO_4)$ replaced by $(OH)_4$, leading to the following formulas: coffinite $= U(SiO4)_{1-x}(OH)_{4x}$ and thorogummite $= Th(SiO_4)_{1-x}(OH)_{4x}$. However, infrared data suggest that coffinite contains molecular water, implying the formula $U(SiO_4) \cdot nH_2O$. The situation is complicated by the fact that most coffinite is extremely fine grained, metamict, possibly impure, and/or altered by groundwater, so analytical work on these minerals remains challenging (Finch and Murakami 1999). Although coffinite never forms attractive specimens for the collector, it is an extremely important component of the Colorado Plateau deposits and is second only to uraninite as a uranium ore mineral.

Thorogummite, like coffinite, is generally very fine grained and metamict. It occasionally forms pseudomorphs after tetragonal thorite crystals, suggesting that the mineral is an alteration or weathering product.

Fig. 53. Thumbnail-sized mass of coffinite from the San Rafael Swell, Emery Co., Utah, showing typical dull black cryptocrystalline habit. *RJL3141*

Huttonite: In the original description (Pabst and Hutton 1951) it was noted that huttonite is rarely metamict, whereas thorite is nearly always so. Huttonite is isostructural with monazite, which is also resistant to metamictization. Thorite is isostructural with zircon, and radioactive zircons are often metamict, leading Pabst to conclude that the huttonite crystal structure is inherently more stable than the thorite structure.

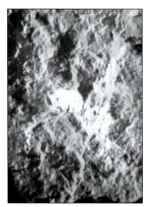

Fig. 54. Scanning electron micrograph of small huttonite grains in matrix from the Holiday mine, Hawthorne, Mineral Co., Nevada. This image was taken in the backscattered electron mode, which causes areas of higher atomic number to appear brighter than areas with lower atomic number, enhancing visual contrast between the huttonite grain in the center and the surrounding silica-rich matrix. *RJL898*

Huttonite was first found as minute colorless grains in sands at Gillespie's Beach (TL) and several other locales in New Zealand. It has been noted at several locales worldwide, generally as small to microscopic transparent colorless adamantine grains or crystals: the Holiday uranium mine, Mineral Co., NV; several deposits in the Leach Lake volcanic complex, Eifel Mts., Germany; Monte Somma, Italy; and others.

# • Secondary Minerals
# Hydrated Oxides

### Hydrated Uranyl Oxides

Secondary hydrated uranyl oxides that do not contain other metal ions include: ianthinite, schoepite, metaschoepite, paraschoepite, studtite, and metastudtite. These minerals tend to form on or very close to uraninite, typically as surface alterations.

Ianthinite, $U^{4+}U^{6+}_5O_{17} \cdot 10H_2O$, is only stable in a somewhat reducing environment, because it contains some $U^{4+}$, which will continue to oxidize under normal atmospheric conditions. It forms minute purple-black laths when fresh; on exposure to air it gradually turns brown and eventually alters completely to yellow schoepite. Ianthinite has been reported from Shinkolobwe, Congo (TL); Wölsendorf, Bavaria; Menzenschwand, Germany; several localities in the Colorado Plateau region; and numerous deposits in France.

Fig. 55. Acicular ianthinite crystals altering pseudomorphously to yellow schoepite from Shinkolobwe, Congo. *RJL686*

Fig. 56. Another view of the specimen in Figure 55 showing the progressive alteration of brown ianthinite to yellow schoepite while retaining the crystal morphology of the iantihinite. *RJL686*

Schoepite, metaschoepite, and paraschoepite are closely related compounds differing only in the water content, with schoepite having the highest hydration. When exposed to air schoepite spontaneously dehydrates, gradually forming a mixture of two or perhaps all three phases. The exact relationship between these species remains unclear and it has been suggested that another phase, "dehydrated schoepite", is also common in nature and that the validity of paraschoepite might be questionable (Finch and Murakami 1999). Schoepite forms golden-yellow, tabular pseudohexagonal crystals and fine-grained crusts. It is fairly abundant at several locales in Congo including Kasolo (TL), Musonoi, and Shinkolobwe; it forms pseudomorphs after uraninite crystals at Beryl Mtn., NH. Some other locales include: Wölsendorf, Germany; Margnac mine, France; and several sites on the Colorado Plateau.

Fig. 57. Earthy, lemon yellow schoepite, with amber becquerelite and orange curite in black uraninite from Shinkolobwe, Congo. This thumbnail-sized sample nicely illustrates the progress of alteration along fractures in the primary uraninite. *RJL353*

Fig. 58. A fascinating pseudomorph: bright yellow schoepite has replaced elongated crystals of demesmaekerite in copper-rich matrix from the Musonoi extension, Shaba Province, Congo. *RJL1699*

Fig. 59. Tabular golden yellow schoepite crystals on aggregates of beige acicular rutherfordine with bright green cuprosklodowskite and dark green malachite, from Musonoi, Congo *RJL2571*

Studtite, $UO_4 \cdot 4H_2O$, and metastudtite, $UO_4 \cdot 2H_2O$, are especially interesting to geochemists because they are the only minerals believed to contain the peroxide (O-O) group. The structural formula for studtite may therefore be expressed as $[(UO_2)(O_2)(H_2O)_2](H_2O)_2$. The compound has been found both on natural uranium and on highly radioactive spent nuclear fuel that had been reacted with water for 1.5 years at 25°C. Attempts to reproduce the compound using uranium oxide containing depleted uranium (i.e., with very little fissile $^{235}U$ isotope) were unsuccessful. It was concluded that the source of peroxide is radiolysis of water by high-energy alpha particles. Thus studtite and metastudtite are the only known minerals that seem to need a radioactive environment in order to form (Burns and Hughes 2003). Studtite was first reported from Shinkolobwe and Kobokobo, Congo, and was considered fairly rare; however, it has since been identified at many localities including Lodeve, France and the Krunkelbach deposit, near Menzenschwand, Germany. It forms aggregates and crusts of minute pale yellow needles associated with other secondary minerals. Metastudtite, also reported from Shinkolobwe, is presumably a dehydration product of studtite.

## Hydrated Uranyl Oxides Containing Other Metal Ions

Hydrated uranyl oxides containing lead are especially important to our understanding of the oxidation zones of geologically old uranium deposits in which substantial amounts of radiogenic lead are present. This group includes curite, fourmarierite, masuyite, richetite, sayrite, spriggite, vandendriesscheite, metavandendriesscheite, and wölsendorfite. These colorful species have great appeal for the collector. Some of the minerals are fairly common, particularly masuyite and vandendriesscheite; others such as richetite and sayrite are quite rare. They are presented here in order of increasing Pb:U ratio, which reflects the process through which the minerals evolve. As more uranium is converted to soluble uranyl ions, which move into the groundwater, the solid phases left behind become increasingly enriched in lead, which is relatively less soluble.

Vandendriesscheite forms yellow-orange to orange pseudohexagonal crystals (up to several mm), subparallel aggregates, and dense microcrystalline intergrowths with metavandendriesscheite. It is the most common member of the group and has been reported from many locales, including: Shinkolobwe, Congo (TL); as a component of *gummite* at the Ruggles pegmatite, NH and the Spruce Pine district, NC; the Eldorado mine, Great Bear Lake, NWT, Canada; Les Bois Noirs, France; and Menzenschwand, Germany. The mineral appears to spontaneously dehydrate in air to some degree, like schoepite, so most if not all specimens likely contain a mixture of vandendriesscheite and metavandendriesscheite.

Fig. 60. Very pale yellow studtite forming bundles of flexible acicular crystals from Menzenschwand, Germany. The small golden yellow crystals are probably billietite. *RJL696*

Fig. 61 Massive primary uraninite with a colorful crust of secondary minerals, including vandendriesscheite (bright orange), fourmarierite (deep reddish orange), and rutherfordine (pale yellow), from Shinkolobwe, Congo. *RJL1144*

Fourmarierite forms reddish-orange to brown tabular pseudohexagonal crystals up to 2 mm across, as well as elongated to acicular crystals and dense crusts. It is a common component of *gummite*, e.g., at Palermo mine, NH and the Spruce Pine district, NC. It is sometimes found in uraniferous mineralized fossil wood, e.g., at Monument No. 2 mine, AZ. Fine specimens are found at Shinkolobwe, Congo (TL); other locales include Eldorado mine, Great Bear Lake, NWT, Canada; Mounana, Gabon; and Wölsendorf, Germany.

Fig. 62. Detail of specimen in Figure 61, showing bright orange vandendriesscheite crystals and pale yellow spherical nodules of rutherfordine. *RJL1144*

Fig. 63. Another area of the same specimen, showing dark orange tabular fourmarierite crystals. *RJL1144*

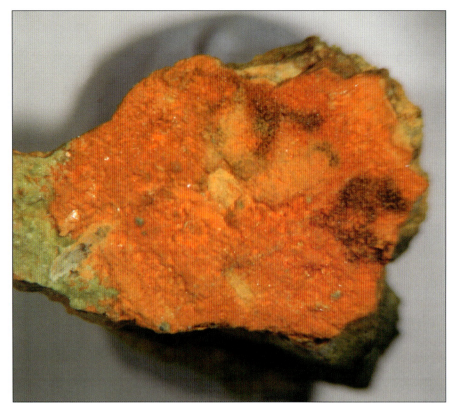

Fig. 64. Fourmarierite forming a bright reddish orange earthy crust on uraninite from Shinkolobwe, Congo. *RJL967*

Richetite is a very rare species, and some details of its formula remain incompletely known. Unlike the other lead-uranyl oxyhydrates, richetite is black to brown, typically forming minute (100 $\mu$m) platy crystals in lamellar aggregates. Known localities are Shinkolobwe, Congo (TL) and Jachymov, Czech Republic.

Masuyite has had a complicated history since it was first described (Vaes 1947) from Shinkolobwe, Congo. Compositions reported by different authors gave Pb:U ratios ranging from 4:9 to 1:3, and from microanalytical studies it was clear that "masuyites" may be zoned, included, or otherwise inhomogeneous. Detailed structural refinements correspond to a Pb:U ratio of 1:3 and a crystal structure closely related to that of protasite (Burns and Hanchar 1999). Masuyite forms tabular pseudohexagonal reddish- to brownish-orange crystals that are always twinned. It is found at Shinkolobwe (TL) and Kamoto, Congo lining cavities in uraninite; other locales include Rabejac, France and the South Alligator Valley, NT, Australia.

Curite forms tiny acicular crystals, microcrystalline to earthy crusts, and dense masses in various shades of reddish-brown to orange. It was first described from Kasolo, Congo associated with torbernite; it is also found at Shinkolobwe. Other locales include: Menzenschwand and Wölsend-orf, Germany; as a component of *gummite* at La Crouzille, France; and a number of occurrences in Canada.

Sayrite is one of the rarest lead uranyl oxyhydroxides. At Shinkolobwe, Congo (TL) it forms small (0.6 mm) yellow-orange to red-orange isolated prismatic crystals associated with uraninite and various secondary uranium minerals.

Fig. 65. Tiny brown plates of richetite, associated with acicular uranophane and other secondary species from Shinkolobwe. Congo. Field of view is about 2mm wide. *RJL676*

Wölsendorfite forms orange-red spherulitic aggregates and crusts of minute crystals associated with other secondary species as an alteration product on or near uraninite. It has been noted at many locales, including: Wölsendorf, Germany (TL); the Eldorado mine, Canada; Shinkolobwe, Congo; the Corrego do Urucum pegmatite, MG, Brazil; the Koongarra uranium deposit, NT, Australia; and several localities in France.

Spriggite is the most recently described member of the group and also the most Pb-rich. It was discovered at the Number 2 workings on Radium Ridge near Mt. Painter, SA, Australia, where it forms bright orange, cm-sized aggregates of randomly oriented prismatic crystals. Individual crystals range up to ~150 $\mu$m long and 40 $\mu$m across. Penetration twins forming six-legged stars are common (Brugger, Ansermet and Pring 2003).

The secondary hydrated uranyl oxides containing alkali, alkaline earth, and transition metals other than lead include: agrinierite, bauranoite, becquerelite, billietite, calciouranoite, metacalciouranoite, clarkeite, compreignacite, protasite, rameauite, uranosphaerite, and vandenbrandeite. Of these, the following three minerals are likely to be of most interest to collectors:

Fig. 66. Veins of orange wölsendorfite with other secondary minerals replacing massive uraninite from Shinkolobwe, Congo. *RJL354*

Becquerelite, $Ca(UO_2)_6O_4(OH)_6 \cdot 8H_2O$, forms tabular to prismatic crystals up to 3 cm long, lathlike crystals, and fine grained crusts in various shades of golden- to lemon-yellow, yellow-orange, and brownish-yellow. It has been identified at many locales, including: Kasolo (TL) and Shinkolobwe, Congo; Wölsendorf, Germany; Eldorado mine, Canada; the Margnac II mine and elsewhere in France; and at numerous sites in the Colorado Plateau region.

Fig. 67. Golden yellow needles of becquerelite on uraninite from Shinkolobwe, Congo. *RJL801*

Fig. 68. Becquerelite crystals forming a solid crust on sandstone from the Happy Jack mine, Utah. *RJL2641*

**54**

Fig. 69. Detail of the becquerelite crystals on the specimen in Figure 68.
*RJL2641*

Billietite, $Ba(UO_2)_6O_4(OH)_6 \cdot 8H_2O$, the barium analogue of becquerelite, forms deep yellow pseudohexagonal tabular crystals to 5 mm that often show twinning. Notable locales include: Shinkolobwe (TL), Swambo, and Luiswishi, Congo; Menzenschwand, Germany; and at Margnac II mine and Rabejac, France.

Fig. 70. Equant, deep amber crystals of billietite from Krunkelbach mine, Menzenschwand, Germany. Specimen is about 3 mm wide.
*RJL582*

Fig. 71. Golden yellow billietite crystals with glassy, pale yellow sprays of acicular sklodowskite from Shinkolobwe, Congo. Field of view is about 4 mm wide. *RJL520*

Fig. 72. Yellow billietite on dark green vandenbrandeite, from Musonoi, Congo. *RJL914*

Vandenbrandeite, $Cu(UO_2)(OH)_4$, forms deep green to nearly black tabular crystals up to 1 cm long. It is found at Kalongwe (TL), Luiswishi, and Musonoi, Congo; also at Rabejac, France.

Fig. 73. Massive, dark green vandenbrandeite with several well-formed mm-sized crystals in a vug, from Musonoi, Congo. *RJL886*

# Carbonates

The secondary uranyl carbonates may be conveniently placed into four groupings: uranyl monocarbonates; uranyl di- and tricarbonates that contain other metal ions; uranyl carbonates that contain rare earths; and other uranyl carbonates, including some that contain other anions.

Uranyl monocarbonates include three species that contain no metals other than U and differ from one another by their degree of hydration: rutherfordine, blatonite, and joliotite.

Rutherfordine often forms the outermost alteration layer on uraninite that is pseudomorphously replaced by *gummite* and uranyl silicates. It is also a component of Colorado Plateau type ores. It is frequently seen at Shinkolobwe and Musonoi, Congo in association with cuprosklodowskite, schoepite, and other secondary minerals. In these specimens the rutherfordine often forms visually distinctive pale yellowish to cream colored bundles of acicular crystals. Other locales include: Morogoro, Tanzania (TL); Palermo No. 3 mine, NH; Newry, ME; the Rabejac deposit, France; and Menzenschwand, Germany.

Uranyl dicarbonates and tricarbonates include andersonite, bayleyite, čejkaite, grimselite, liebigite, rabbittite, swartzite, widenmannite, zellerite, metazellerite, and znucalite.

Fig. 74. Close-up of rutherfordine in a cuprosklodowskite nodule from Musonoi, Congo, showing the typical habit: pale cream-colored bundles of acicular crystals. Bright yellow crystals are schoepite. *RJL2571*

Andersonite is water soluble and is usually found as an efflorescence on mine walls in fairly arid settings. It is strongly fluorescent bright green under UV light. At the Hillside mine, AZ (TL) minute clusters of pseudocubic crystals were associated with gypsum, bayleyite, schroekingerite, and swartzite. At the Atomic King No. 2 mine, near Moab, UT thick crusts of andersonite were mined as ore and crystals up to 1 cm were found. Some other locales include: Jim Thorpe, PA; Jachymov, Czech Republic; Ronneburg, Germany; and the Geevor mine, Cornwall, England.

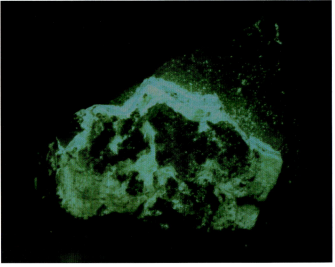

Fig. 76. The andersonite specimen from Figure 75, shown here in SW UV light. *RJL767*

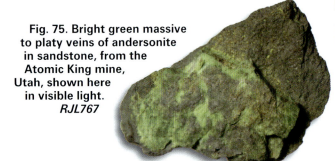

Fig. 75. Bright green massive to platy veins of andersonite in sandstone, from the Atomic King mine, Utah, shown here in visible light. *RJL767*

Fig. 77. Scanning electron micrograph of andersonite from the Atomic King mine, Utah, showing platy crystals in a parallel arrangement. Field of view is about 1 mm wide. *RJL767*

Bayleyite is also water soluble and occurs with andersonite as an efflorescence at the Hillside mine, AZ (TL) where it forms yellow crusts and small prismatic crystals that are weakly fluorescent under UV.

Other locales include: the Pumpkin Buttes area of the Powder River Basin, WY; the Ambrosia Lake district, NM; numerous mines in UT; and a mine tunnel at Aznegour, Morocco.

Fig. 78. Light green 1-2 mm bayleyite crystals scattered on matrix from the Hillside mine, Yavapai Co., Arizona. *RJL1376*

Fig. 79. Detail of the bayleyite crystals in Figure 74. *RJL1376*

Liebigite usually forms bright green to yellowish-green granular to scaly crusts and films; when it does form distinct crystals they are usually crude, with rounded edges and a waxy appearance. Good specimens are colorful in natural light and show brilliant bluish-green fluorescence under UV light. The classic American locality for liebigite is the Schwartzwalder mine, Jefferson Co., CO, but it is fairly widespread throughout the western U.S., including: the Mi Vida mine, UT; the Pumpkin Buttes area, WY; the Lucky Mc mine, WY; the Hanosh mines, near Grants, NM; and the Midnite mine, WA. Other locales include: Jachymov, Czech Republic, where it is plentiful in the oxidized zone of uraninite veins and as an efflorescence on mine walls; Schneeberg, Germany; and Wheal Basset, Cornwall, England.

Uranyl carbonates containing essential REE include astrocyanite-(Ce), bijvoetite-(Y), kamotoite-(Y), lepersonnite-(Gd), and shabaite-(Nd). Of these, bijvoetite-(Y) and lepersonnite-(Gd) were described from the oxidation zone at Shinkolobwe, Congo (Deliens et al. 1984). The other three species were discovered during work at the Kamoto East extension, Congo (Deliens et al. 1990).

Fig. 80. Waxy yellow liebigite thickly covering schist from the Schwartzwalder mine, Boulder, Colorado, shown here in visible light. *RJL885*

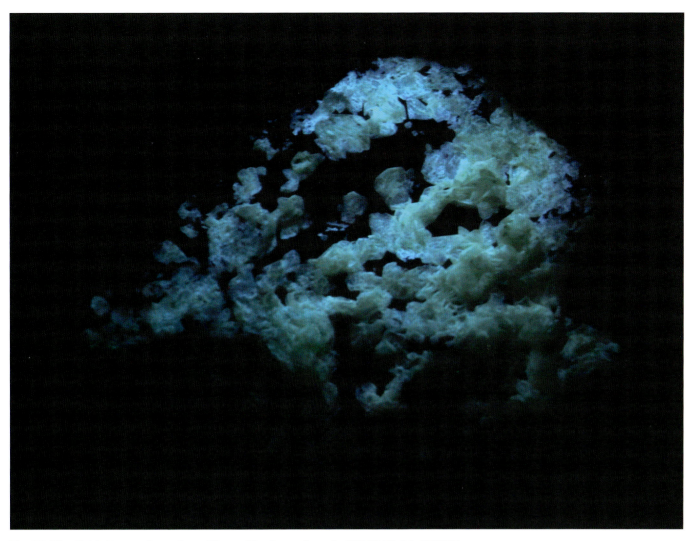

Fig. 81. The liebigite specimen from Figure 80, shown here in SW UV light. *RJL885*

Fig. 82. Liebigite crystals to about 2 mm from the Pitch mine, Saguache Co., Colorado. *RJL3063*

Kamotoite-(Y) is a monoclinic hydrous rare earth uranyl carbonate; the type material contains several percent each of Nd, Sm, Gd, and Dy in partial substitution for Y. Two somewhat distinct habits are seen: transparent, bright yellow bladed crystals to about 5 mm, and flattened fanlike aggregates. The mineral effervesces strongly in dilute HCl, and the effervescence and crystal habit help to distinguish it from other yellow secondary minerals that occur at Kamoto East. Dehydration causes the crystals to become opaque and the color becomes less intense. Kamotoite-(Y) is one of the more abundant secondary minerals in uraniferous pockets at Kamoto East. It can form cm-sized layers on altered uraninite in the supergene zone. Associated minerals include pale yellow rosettes of shabaite-(Nd), minute acicular crystals and spherical nodules of uranophane, blue rosettes of astrocyanite-(Ce), and transparent blue-green crystals of schuilingite-(Nd). It has also been found sparingly at the Gole quarry, near Madawaska, ONT, Canada.

Other uranyl carbonates include albrechtschraufite, fontanite, schroeckingerite, sharpite, urancalcarite, voglite, and wyartite. Of these, wyartite is notable as the first mineral having pentavalent uranium ($U^{5+}$) in its structure (Burns and Finch 1999a).

Fig. 83. Bright yellow kamotoite-(Y) showing the typical habit of radiating flattened opaque crystals in a fanlike arrangement, from the type locale, Kamoto East, Congo. RJL1372

Fig. 84. Another specimen of kamotoite-(Y) from the type locale illustrating an alternate, less common habit: tiny transparent golden yellow crystals on altered uraninite. RJL2569

Schroeckingerite is a hydrated sodium calcium uranyl fluo-carbonate-sulfate that is soluble in water and dilute acids. It is pale greenish-yellow with vitreous to pearly luster and shows bright yellow-green fluorescence under UV light. Distinct crystals are rare; the common habit is crusts and globular aggregates of minute scaly grains. The water content can vary zeolitically to some degree with humidity changes. Schroeckingerite is plentiful along Lost Creek, near Wamsutter, Sweetwater Co., WY, where it forms small concretionary aggregates in clays and silts containing gypsum. The deposit is a very recent near surface caliche-type formation (Frondel 1958). The mineral is widespread at sites in the western U.S., including: the Hillside mine, AZ; the Poison Basin area, Carbon Co., WY; and numerous districts in UT. Worldwide occurrences include: Jachymov, Czech Republic (TL); Johanngeorgenstadt, Germany; the La Soberania mine, Mendoza Province, Argentina; and the Zeehan district, Tasmania, Australia.

**Fig. 85. Pale yellow pearly schroeckingerite forming a pea-sized nodule in clay from Sweetwater Co, WY, shown here in natural light.** *RJL570*

**Fig. 86. The sample of schroekingerite in Figure 85, shown in SW UV light.** *RJL570*

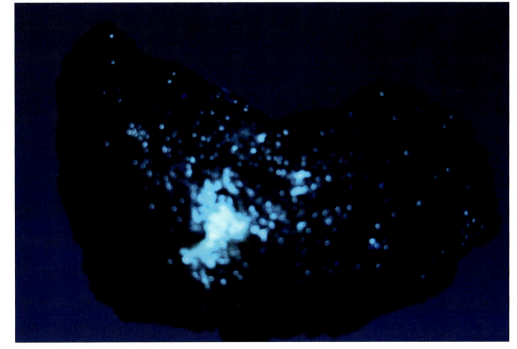

Fig. 87. Pale yellow plates and scales of schroeckingerite on a sulfide-rich matrix from the Zeehan mining district, Tasmania, shown here in natural light. *RJL689*

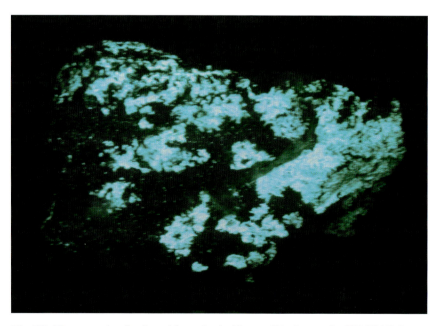

Fig. 88. The sample of schroekingerite in Figure 87, shown in SW UV light. *RJL689*

Voglite, a hydrated calcium copper uranyl carbonate, forms emerald- to grass-green aggregates of microscopic scaly crystals. It resembles liebigite, with which it is sometimes found, but voglite is not fluorescent. It is comparatively rare; documented localities include: the Elias mine, Jachymov, Czech Republic (TL); Frey Point and the White Canyon No. 1 mine, San Juan Co., UT; and the Red Mesa district, Navajo Co., AZ.

Fig. 89. Platy, emerald-green voglite forming a seam about 1 mm thick in matrix from the White Canyon district, San Juan Co., Utah. *RJL756*

Fig. 90. Layer of yellow liebigite about 3 mm thick growing on a thin seam of green voglite from the White Canyon district, San Juan Co., Utah. *RJL755*

# Sulfates

## The Zippeite Group

The minerals of the zippeite group are monoclinic or triclinic uranyl sulfates also containing alkali or transition metals. For alkalis, the general formula may be expressed as: $A_x^{+1}(H_2O)_y[(UO_2)_a(SO_4)_bO_c(OH)_d]$, where $A = K$ or $Na$. For transition metals, the general formula is $B^{+2}(H_2O)_{3.5}[(UO_2)_2(SO_4)O_2]$, where $B = Co$, $Mg$, $Ni$, or $Zn$.

The first naturally-occurring uranium sulfates were studied by J. F. John in 1821, who analyzed an emerald-green "uranvitriol" and a yellow earthy "basisches schwefelsaueres uranoxyd", both from Jachymov, Czech Republic. The green mineral was formally described in 1830 as johannite by Haidinger, who provided a crystallographic analysis. Haidinger later applied the name zippeite to the yellow earthy material, but did not provide descriptive data.

Members of the zippeite group are fairly widespread but for many years they remained poorly understood, partly because the mineral is commonly extremely fine powder with admixed impurities, and partly because the minerals were long believed to be simply hydrated uranyl sulfates. Detailed studies of synthetic zippeite-like phases showed that potassium was an essential constituent, and re-analysis of natural materials confirmed that zippeite is actually a potassium uranyl sulfate. Tandem studies of synthetic and natural materials led to the formal recognition of sodium-zippeite, cobalt-zippeite, magnesium-zippeite, nickel-zippeite, and zinc-zippeite (Frondel et al. 1976).

Fig. 91. Hand specimen with pinhead-sized bright yellow earthy to microcrystalline masses of zippeite on massive black coffinite from the Sec. 35 mine, Grants, New Mexico. *RJL3050*

Fig. 92. In the microscope, colorless gypsum crystals can be seen among the yellow zippeite aggregates in this sample from the Sec. 35 mine, Grants, New Mexico. *RJL346*

All of these species are secondary minerals that form in the oxidized zone of uraninite deposits that contain sulfide minerals. They often form efflorescences on mine walls, and are commonly associated with gypsum, johannite, and uranopilite. Sodium-zippeite is typical of sandstone-type deposits, whereas zippeite is more characteristic of weathered uraninite veins in igneous or metamorphic rocks (Frondel et al. 1976). Members of the group cannot be distinguished from one another reliably by visual examination, and many "zippeite" specimens contain more than one member of the group in close association.

Fig. 93. Scanning electron micrograph showing individual zippeite crystals from Grants, New Mexico. Field of view is about 50 $\mu$m wide. *RJL750*

Fig. 94. Bright yellow velvety crust of sodium zippeite on matrix from a recent find at Green River, Emery Co., Utah. *RJL2662*

Fig. 95. Microscopic orange aggregates of marecottite from the type locale, La Creusaz, near Les Marecottes, Switzerland. *RJL3058*

Fig. 96. Two tiny aggregates of pseudojohannite illustrating lichen-like habit, from the type locale, Jachymov, Czech Republic. *RJL3067*

Other secondary uranyl sulfates include: deliensite, jachymovite, johannite, rabejacite, and uranopilite. Several authors have described specimens under the name *meta-uranopilite*, but this species remains poorly studied and its validity is dubious.

Fig. 97. Pale green microcrystalline johannite with yellow earthy zippeite from Roughton Consols, Bodmin Moor, Cornwall, England. *RJL1014*

Fig. 98. Minute acicular yellow crystals of uranopilite forming rounded velvety aggregates from Les Bois Noirs, Loire, France. *RJL794*

# Selenites, Tellurites, and Arsenites

Uranyl selenite minerals form where selenium-bearing sulfide minerals are undergoing oxidation and dissolution; in all these minerals, selenium is in the tetravalent state as the selenite ion, $(SeO_3)^{-2}$. Most uranyl selenites are found at the Musonoi deposit in the Congo, where the source of selenium is the oxidation of selenian digenite (Deliens et al. 1981). The minerals in this group are: demesmaekerite, derriksite, guilleminite, haynesite, larisaite, marthozite, and piretite.

Demesmaekerite is one of several distinctive uranyl selenite minerals originally described from the Musonoi mine, Congo. It forms triclinic crystals that are bottle green when fresh; with dehydration the color darkens to olive green or brownish and the material becomes more opaque. It is not fluorescent in UV. It forms in the lower part of the oxidation zone at the Musonoi mine, Kolwezi, Congo in association with cuprosklodowskite, kasolite, malachite, derricksite, guilliminite, and chalcomenite in selenian digenite. Pseudomorphs of schoepite after demesmaekerite have been found at Musonoi.

Fig. 99. Small crystals of demesmaekerite from the type locale, Musonoi, Congo, showing progressive alteration. Fresh crystals are very dark transparent green; with alteration they become lighter green and opaque. *RJL815*

Derricksite is one of the rarer of the uranyl selenites in the Musonoi deposit. It forms minute euhedral crystals less than about 0.7 mm across, in colors ranging from clear green to malachite green and rarely bottle green. It is not fluorescent. Derricksite forms in the lower part of the oxidation zone at the Musonoi mine, Kolwezi, Congo in association with dark green demesmaekerite, acicular yellow guilleminite, green to yellow marthozite, and blue chalcomenite. As with the associates, derricksite is an alteration product of primary uraninite and selenian digenite.

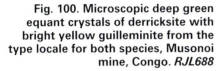

Fig. 100. Microscopic deep green equant crystals of derricksite with bright yellow guilleminite from the type locale for both species, Musonoi mine, Congo. *RJL688*

Guilleminite is a barium uranyl selenite found in the Musonoi deposit. It forms canary-yellow powdery to microcrystalline coatings, silky to fibrous masses, and small spherical nodules in cavities in the host rock. The crystals are typically opaque to translucent, rarely transparent in very small crystals. Guilleminite is not fluorescent. It occurs in the oxidation zone at the Musonoi mine, Kolwezi, Congo in association with malachite, wulfenite and other secondary uranium minerals, including: uranophane, kasolite, vandenbrandeite, curposklodowskite, and metatorbernite. It is frequently associated with the other uranyl selenites. It has also been found occasionally at Shinkolobwe, where it is associated with hydrated uranium oxides, rutherfordine, and uranophane.

Fig. 101. Bright yellow earthy guilleminite on microcrystalline malachite from Musonoi, Congo. *RJL341*

Fig. 102. Bright yellow guilleminite covering a small patch of olive-green marthozite from Musonoi, Congo. *RJL1140*

Haynesite was named in 1991 based on materials collected by Patrick Haynes at the Repete Mine, San Juan Co., UT. It forms transparent amber-yellow vitreous crystals, as elongated tablets or acicular prismatic rosettes to several mm in diameter. When the crystals are very small, the material forms yellowish smears and tiny felted masses; microscopic examination at 20X reveals minute, pale yellow fibrous crystals. At the larger sizes, the crystals are deep golden yellow and are clearly visible to the naked eye. Associated minerals at the type locality include andersonite, boltwoodite, larisaite, ferroselite, uraninite, marcasite, and gypsum.

Fig. 103. Golden yellow needles of haynesite on matrix from the type locale, the Repete mine, San Juan Co., Utah. *RJL2566*

Fig. 104. Detail of specimen shown in Figure 103, showing divergent sprays of transparent yellow haynesite crystals. *RJL2566*

Marthozite is a copper uranyl selenite that forms distinctive triangular or truncated pyramidal crystals that are flattened and typically striated. The color ranges from yellowish green to greenish brown. Crystals are transparent when fresh, becoming opaque upon dehydration. It is not fluorescent. Marthozite forms in the lower part of the oxidation zone at the Musonoi mine, Kolwezi, Congo in association with cuprosklodowskite, kasolite, malachite, derricksite, demesmaekerite, guilleminite, sengierite, and chalcomenite.

Fig. 105. Platy, yellowish-green marthozite on selenian digenite from the type locale, Musonoi, Congo. *RJL1145*

All three of the known uranyl tellurite minerals, cliffordite, schmitterite, and moctezumite, are found at the San Miguel and Moctezuma mines in Moctezuma, Sonora, Mexico. One of them, cliffordite, also occurs at Shinkolobwe, Congo. These minerals form where sulfide minerals are undergoing oxidation and the Te is presumably an impurity in the sulfides. In contrast to the uranyl selenites, all of the uranyl tellurites are anhydrous.

An anhydrous uranyl arsenite mineral, chadwickite, was described from a mine in the central Black Forest, Germany. It does not appear to have any close relationship to other species.

Fig. 106. Yellow microcrystalline cliffordite with globular aggregates of dark green mackayite from the type locale, Moctezuma, Sonora, Mexico. *RJL338*

Fig. 107. Bright orange microcrystals and crystalline films of moctezumite from the type locale, Moctezuma, Sonora, Mexico. *RJL2802*

# Molybdates and Tungstates

Uranyl molybdates are important ore minerals at some localities and are also common accessory minerals in roll-front deposits and other settings where both uraninite and Mo-bearing species are being weathered. In these minerals the Mo is present as $Mo^{+6}$ and the U may be of mixed valence, $U^{+4}$ and $U^{+6}$. It is likely that these minerals are more abundant than commonly believed, because of the fact that they are often difficult to distinguish in the field and are often intergrown with one another in fine-grained masses (Finch and Murakami 1999). The members of this group are of limited interest to collectors because they don't make attractive specimens and cannot be reliably identified without advanced analytical facilities. The uranyl molybdates include calcurmolite, cousinite, deloryite, iriginite, moluranite, tengchongite, and umohoite. There is only one uranyl tungstate, uranotungstite, reported from several localities in Germany.

## Phosphates and Arsenates

There are over seventy uranyl phosphates and arsenates, many of which are widely distributed in nature and of particular interest to collectors. To provide some order to this large family of minerals, it is convenient to place them into several groupings: the autunite group; the phosphuranylite group; a group structurally related to walpurgite; and other phosphates and arsenates that do not seem to be closely related to the foregoing groups (Finch and Murakami 1999).

### The Autunite Group

The autunite group is divided into two subgroups, based on crystal structure differences related to the degree of hydration. For each cation species there are usually several hydration states, with the lower hydration being generally more stable at room temperature and ordinary humidity. When there is a single lower hydration state it is designated by the prefix "meta-" (e.g., torbernite and metatorbernite). When many stable hydration states are observed, they may be designated by Roman numerals (e.g., "meta-uranocircite I" and "meta-uranocircite II"). For a collector devoted to accurate labeling, this presents a dilemma. Should a partially dehydrated sample be labeled "autunite", "autunite with meta-autunite" or "meta-autunite pseudomorph after autunite"? It is the author's view that the real value of a mineral specimen lies in its information content. What can the rock tell us about its origin? The original mineral assemblage can provide valuable insights into the thermodynamic conditions that were present at the time of formation, and labeling practice should recognize that this information is more important in most cases than artifacts that might have occurred later during storage. The label should therefore reflect (to the extent known) the species that was originally formed.

The autunite subgroup minerals are tetragonal uranyl phosphates, arsenates, and vanadates with the general formula $A(UO_2)_3(XO_4)_2 \cdot 8\text{-}12H_2O$, where A = Ba, Ca, Cu, $Fe^{+2}$, ½(HAl), Mg, $Mn^{+2}$, or $Na_3(UO_2)$, and X = As, P, or V.

**Fig. 108. Dull yellow earthy iriginite from Ben Lomand, Queensland, Australia.** *RJL1349*

**Fig. 109. Waxy yellow film of iriginite from the type locale, Aleksandrovskiy Golets Mo-U deposit, Transbaikal, Russia.** *RJL3145*

The subgroup includes autunite, heinrichite, kahlerite, novačekite, uranocircite, sabugalite, saleeite, torbernite, trogerite, uranospinite, xiangjianite, and zeunerite.

The meta-autunite subgroup minerals are tetragonal or orthorhombic uranyl phosphates and arsenates with the general formula $A(UO_2)_3(XO_4)_2 \cdot 4\text{-}8H_2O$, where A = Ba, Ca, Co, Cu, $Fe^{+2}$, $(H_3O)_2$, $K_2$, Mg, $(NH_4)_2$, or Zn, and X = $As^{+5}$ or $P^{+5}$.

The subgroup includes abernathyite, bassetite, chernikovite, lehnerite, meta-ankoleite, meta-autunite, metaheinrichite, metakahlerite, metakirchheimerite, metalodevite, metanatroautunite, sodium uranospinite, metanovačekite, metatorbernite, meta-uranocircite, meta-uranospinite, metazeunerite, przhevalskite, and uramphite.

Members of both subgroups are generally quite similar in their habits and outward appearance, forming euhedral, generally tabular crystals in shades of yellow to green. Most are strongly fluorescent under UV light; the notable exceptions are the copper-containing species torbernite, metatorbernite, zeunerite, and metazeunerite. Some members are widely distributed: autunite has been documented at over 800 locales and torbernite at over 600. Others are fairly rare, e.g., uramphite, which is known from a single locality.

Fig. 110. Autunite crystals in cavernous matrix from Hellemann-Stollen, Germany. *RJL973*

Fig. 111. A thumbnail-sized cluster of autunite crystals from the São Pedro mine, Brazil. *RJL1559*

Fig. 112. A particularly aesthetic autunite specimen from Luzy, France. *RJL663*

Fig. 113. Solid vein of dark green autunite altering to lighter meta-autunite at edges, from the Daybreak mine, Washington. *RJL19*

Fig. 114. Pearly scales of heinrichite on matrix from the type locale, the White King mine, Oregon. *RJL569*

Fig. 115. A small novačekite crystal group from Brumado, Bahia, Brazil. *RJL1378*

Fig. 116. Straw-yellow saleeite crystals thickly scattered on matrix from Rum Jungle, Northern Territory, Australia. *RJL571*

Fig. 117. A classic specimen of torbernite from Gunnislake mine, Cornwall, England. Main crystal group is about 2 cm across. *RJL769*

Fig. 118. Torbernite crystals lining a cavity in ore from Shaba province, Congo. *RJL623*

Fig. 119. Pearly green torbernite crystals on matrix from La Luz, Sonora, Mexico. *RJL2860*

Fig. 120. A closer view of the uranocircite from Streuberg, Saxony, Germany shown earlier in Figure 29. *RJL3112*

Fig. 121. Transparent green zeunerite microcrystals on blue drusy quartz from the Copper Stope, Majuba Hill, Nevada. *RJL909*

Fig. 122. Pearly green metazeunerite microcrystals lining a narrow fissure from the Copper Stope, Majuba Hill, Nevada. *RJL782*

Fig. 123. Thin tabular meta-autunite crystals in subparallel clusters forming a solid mass from Sternbruch, Germany. *RJL2625*

Fig. 124. Metatorbernite crystals to about 2 mm from Cuna Baixa mine, Viseu, Portugal. Note the color zoning in the crystals at center. *RJL347*

Fig. 125. Cluster of pearly green meta-torbernite crystals from Trancoso, Portugal. *RJL768*

Fig. 126. Metatorbernite crystals filling cavity in matrix from Shinkolobwe, Congo. *RJL2570*

Fig. 127. Chartreuse plates of meta-uranocircite from the type locale, Bergen, Germany. *RJL777*

Fig. 128. Metazeunerite crystals to 5 mm scattered on matrix from Cinovec (Zinnwald) Czech Republic. *RJL2381*

Fig. 129. Plates and flakes of metazeunerite to about 2 mm scattered on friable sandstone from Adelaide River, Northern Territory, Australia. *RJL572*

Fig. 130. Equant crystals of metazeunerite from the Erongo Mts., Namibia. *RJL 2783*

## The Phosphuranylite Group

The phosphuranylite group is based on a structural sheet with U:P = 3:2 suggesting the general formula $(M^{p+})_m[(UO_2)_3(O,OH)_2(PO_4)_2]_p \cdot _{m/2}(H_2O)_n$. This concept of the group includes: althupite, arsenuranylite, bergenite, dewindtite, dumontite, francoisite-(Nd), hügelite, kivuite, mundite, phosphuranylite, phuralumite, phurcalite, renardite, upalite, vanmeersscheite, metavanmeersscheite, and yingjiangite. (Many authors consider renardite to be identical to dewindtite. Kivuite is also regarded by some as a doubtful species.)

Fig. 131. Minute yellow tabular dewindtite crystals with orange curite and other secondary minerals from the type locale, Shinkolobwe, Congo. *RJL352*

Fig. 132. Microcrystalline dewindtite forming a yellow crust on matrix with minor green torbernite from the La Faye mine, Saone-et-Loire, France. *RJL575*

Fig. 133. Radiating clusters of tiny brown tabular dewindtite crystals on torbernite from the La Faye mine, Saone-et-Loire, France. *RJL1030*

Fig. 134. Pale yellow dumontite crystals from
Shinkolobwe, Congo. Longest crystals are about
1 mm. *RJL340*

Fig. 135. Hand specimen of phurcalite and malachite on friable red sandstone from Posey mine,
Utah. *RJL2643*

Fig. 136. A closer view of the specimen shown in Figure 135, showing radiating spherical aggregates of the two minerals in intimate association. *RJL2643*

Phosphuranylite, a calcium uranyl phosphate, typically forms earthy to scaly crusts of minute deep golden yellow crystals. Distinct crystals, when present, are generally lathlike with a rectangular outline. It is not fluorescent. Phosphuranylite is widely distributed (well over 100 locales) but usually found in relatively small amounts. The most common mode of occurrence is an alteration product of uraninite in radioactive pegmatites, e.g., at the Flat Rock pegmatite, Mitchell Co., NC (TL); the Ruggles and Palermo pegmatites, Grafton Co., NH; Bedford, NY; and Keystone, SD. It has been noted at several important uranium ore deposits, including: Bergen on the Trieb and Wölsendorf, Germany; La Crouzille and Margnac, France; and numerous locales in the Colorado Plateau region.

Fig. 137. Bright yellow crust of phosphuranylite lining a thin fissure in ore from Margnac, mine, Haut-Vienne, France. Interestingly, the phosphuranylite formed on top of a pale yellow layer that is strongly fluorescent, probably a member of the autunite group. RJL573

---

Bergenite, the barium analogue of phosphuranylite, forms clusters and coatings of bright yellow thin tabular crystals. It is weakly fluorescent pale green under UV light according to several recent handbooks, although the original description noted "weak orange-brown" fluorescence (Bultemann and Moh 1959). Analyzed material from this author's research collection shows weak green fluorescence under SW UV light. Bergenite was originally found on a mine dump at Streuberg, near Bergen on the Trieb, Vogtland, Germany (TL) where it was associated with uranocircite, torbernite, renardite, autunite, and other secondary uranium minerals. The original locale has since been reclaimed. Bergenite is also found at Menzenschwand, Germany and in graphitic uranium ore of the Black Slate Formation, South Korea.

Fig. 138. A closer view of the specimen shown in Figure 28, showing a 1 cm wide mass of platy yellow bergenite in a quartz lined cavity from Mechelgrun, Saxony,Germany. *RJL2248*

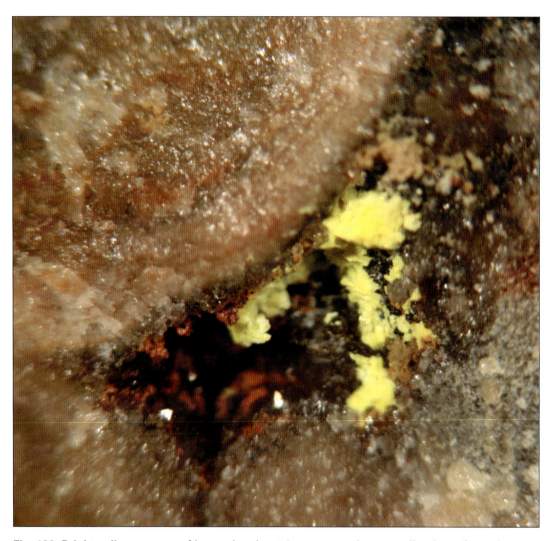

Fig. 139. Bright yellow masses of bergenite about 1 mm across in quartz-lined vug from the type locale, Streuberg, Bergen, Germany. *RJL703*

Phosphates and arsenates related to walpurgite have U:P and U:As ratios of 1:2 suggesting the general formula $M^{2+}[(UO_2)(PO_4)_2](H_2O)_n$. This concept of the group includes: hallimondite, orthowalpurgite, parsonsite, pseudoautunite, ulrichite, and walpurgite.

Parsonsite forms submillimeter lathlike crystals, often in radial clusters and tufts or as earthy crusts and radial-fibrous aggregates. Color ranges from pale yellow to honey-brown; earthy forms are darker brown due to admixed impurities. It is not fluorescent. Parsonsite is found in many localities worldwide, and at Lachaux, France it is a significant component of the ore. Other occurrences include: Kasolo, Congo (TL), associated with kasolite, torbernite, and dewindtite; with meta-uranocircite at Wölsendorf, Germany; the Ruggles pegmatite, Grafton, NH, on fracture surfaces near uraninite and *gummite* masses; and the Ranger mine, NT, Australia.

Fig. 140. Minute honey colored tabular crystals of walpurgite from Schmiedestollen mine, Germany. Field of view is about 2 mm wide. *RJL835*

Fig. 141. A sample of ore from Grury, Saone-et-Loire, France. Parsonsite is the main uranium mineral, richly coating surfaces in a tuffaceous matrix. *RJL2647*

Fig. 142. Detail of another sample from Grury, similar to the one shown in Figure 141, showing a thick coating of microscopic pale yellow parsonsite crystals. *RJL2646*

Ulrichite is a very rare calcium copper uranyl phosphate that forms divergent sprays of bright green acicular crystals in miarolytic cavities in granite. It is thought to have crystallized from late-stage hydrothermal solutions and is associated with a number of other Cu-containing species including chalcosiderite, libethenite, turquoise, pseudomalachite, and torbernite. Ulrichite is known from a single locality, the Lake Boga granite, near Swan Hill, Victoria, Australia (Birch et al. 1988). Despite its limited distribution, the availability of small samples of ulrichite for collectors is fairly good as of this writing.

Other uranyl phosphates and arsenates include arsenuranospathite, asselbornite, chistyakovaite, coconinoite, furongite, kamitugaite, moreauite, ranunculite, seelite, threadgoldite, triangulite, uranospathite, and vochtenite.

Fig. 143. Detail of the specimen shown earlier in Figure 38, showing a small spray of green ulrichite in granite from the type locale, Lake Boga, Australia. *RJL1330*

# Vanadates

## The Carnotite Group

Minerals of the carnotite group are orthorhombic or monoclinic uranyl vanadates that are chemically related to the autunite group minerals but have a distinctive crystal structure. The minerals have the general formula: $A(UO_2)_2V_2O_8 \cdot nH_2O$, where A = Mn, Ca, Ba, Pb, $K_2$, $Cs_2$, $Na_2$, or $H_3O$, and n = 1 to 6.

The group includes carnotite, curienite, francevillite, fritzschite, margaritasite, strelkinite, tyuyamunite, metatyuyamunite, vanuralite, metavanuralite, and vanuranylite.

Fig. 144. Yellow powdery curienite on orange francevillite from the type locale for both minerals, Mounana, Gabon. *RJL903*

*96*

Fig. 145. A closer view of the specimen shown in Figure 144, illustrating the radiating crust of francevillite with bright yellow curienite sprinkled about the surface. *RJL903*

Fig. 146. Scanning electron micrograph showing minute tabular curienite crystals from a sample very similar to that shown in Figures 144-145, from Mounana, Gabon. Field of view is about 200 μm wide. *RJL899*

Fig. 147. Golden yellow mass of vanuralite about 1 cm tall from Lavra Jabuti, Baixio, Minas Gerais, Brazil. *RJL2889*

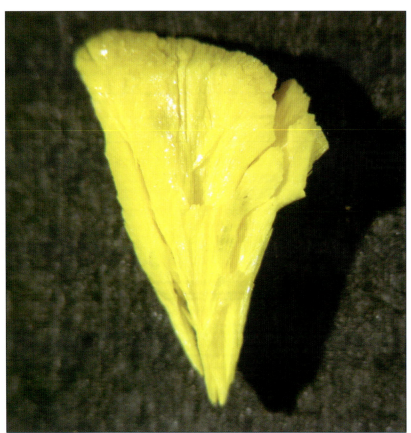

Fig. 148. Pure radiating mass of vanuralite about 1 cm tall from the type locale, Mounana, Gabon. *RJL470*

Carnotite forms microscopic flattened rhomboidal crystals, as microcrystalline aggregates, crusts, and earthy masses. It is bright yellow to greenish-yellow; when distinct crystals are present they have a pearly to waxy luster. Carnotite is an important component of sandstone-type ores and is widespread throughout the western U.S.; it is also the main ore mineral at the Tyuya Muyun deposit, Kyrgyzstan. Distinct crystals on malachite are found at Mashamba-West, Congo; at Radium Hill, near Olary, SA, Australia it forms surface alterations on davidite-(La).

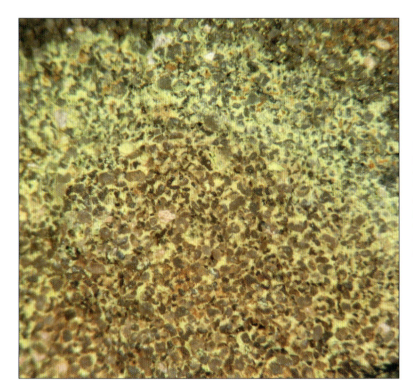

Fig. 149. Optical microscope photo of earthy yellow carnotite impregnating sandstone ore from Grants, New Mexico. One can see how this type of ore can be exploited by *in-situ* leaching, in which a chemical solution is pumped into the sandstone layer to simply dissolve the uranium minerals and yield a uranium-rich solution to be pumped out by an adjoining well. *RJL69*

Fig. 150. Butterscotch-colored microcrystals of carnotite from the Anderson mine, Yavapai Co., Arizona. *RJL226*

Francevillite forms tabular crystals, often in stacked or subparallel aggregates, in various shades of yellow-orange, greenish yellow, green, and brown. At the Mounana mine, Franceville, Gabon (TL) brilliant orange crystals scattered on dark brown mounanaite make spectacularly colorful specimens. Small but excellent yellow-green crystals on black sandstone are found at the Pandora mine, San Juan Co., UT. Other locales include the South Terras mine, Cornwall, England; Nussbach, Germany; and the Oktyabrskoe deposit, near Krasnokamensk, Siberia, Russia.

Fig. 151. A closer view of francevillite on mounanaite from Mounana, Gabon, shown earlier in Figure 42, showing the fanlike aggregates of francevillite distributed on a layer of darker brown mounanaite. *RJL1353*

Fig. 152. Scanning electron micrograph of another sample of francevillite from Mounana, Gabon, showing the crystal morphology in greater detail. Field of view is about 0.5 mm wide. *RJL746*

Fig. 153. Tiny, olive-green clusters of francevillite crystals from the Pandora mine, San Juan Co., Utah. *RJL2642*

Margaritasite, the cesium analogue of carnotite, forms fine-grained aggregates of 1 to 3 μm tabular yellow crystals. The powdery aggregates typically fill tiny pores and relict phenocryst casts in a rhyodacite breccia of reworked tuff clasts in the lower Escuadra Formation at the Margaritas #2 uranium deposit, Sierra Peña Blanca, west of Chihuahua City, Mexico (TL). Associated minerals at the Margaritas deposit include uranophane, uranophane-ß, autunite, and weeksite. Geochemical data suggest that margaritasite is the product of local hydrothermal or pneumatolitic activity during or after uranium mineralization. It would thus be unlikely to find margaritasite in any Colorado Plateau type deposit; conversely, in uranium deposits of probable hydrothermal origin, any "carnotite" present should be examined as a potential new occurrence of margaritasite (Wenrich et al. 1982).

Fig. 154. A small bleb of yellow earthy margaritasite lining a small pore in matrix from the type locale, Margaritas #2, Sierra Peña Blanca, Mexico. *RJL418*

Tyuyamunite, the calcium analogue of carnotite, has similar properties and modes of occurrence. Crystals are tiny bright yellow flattened scales and laths, usually in fanlike or radial aggregates or as an earthy component filling sandstone type deposits. Fluorescence under UV light is either very weak yellow-green or absent altogether. It is somewhat more likely than carnotite to form distinct crystals. Notable localities include: Tyuya Muyun, Ferghana Valley, Kyrgyzstan (TL); numerous sites in the Colorado Plateau region; as fine specimens from the Marie and Dandy mines, Pryor Mtns. district, Carbon Co., MT; Rozmital, Czech Republic; Dortmorsbach, Germany; and Musonoi, Congo.

Other uranyl vanadates include ferghanite, rauvite, and uvanite. Several of these species are poorly defined and further work is needed to elucidate their structures and their relationship to the carnotite-group minerals. Ferghanite might be closely related to or identical with metatyuyamunite.

Fig. 155. Divergent sprays of golden yellow tyuyamunite crystals thickly lining a 2 cm cavity, associated with dark green fibrous malachite and colorless calcite rhombs from Musonoi, Congo. *RJL797*

Fig. 156. Scanning electron micrograph of a small tyuyamunite crystal group removed from the sample in Figure 155, showing details of the crystal morphology. *RJL797*

Fig. 157. Bright yellow crystals of tyuyamunite on sandstone from the Marie mine, Pryor Mts. district, Montana. *RJL2584*

Fig. 158. Detail of another specimen of tyuyamunite from the Marie mine, Montana, showing small rosettes of pearly yellow tabular crystals. *RJL2585*

Fig. 159. Another type of sample from the Pryor Mts. district: a plate of honey colored scalenohedral calcite crystals coated with a layer of microcrystalline tyuyamunite from the Dandy mine, Montana. *RJL3055*

# Silicates

The uranyl silicates are the most abundant uranyl minerals, owing to their relatively low solubility in most groundwaters and the relative ubiquity of dissolved Si in those waters. For this reason these minerals are also fairly stable and rarely alter to other species once formed. The minerals may be divided into several groups based on the atom ratio of uranium to silicon. The 1:1 uranyl silicates comprise the uranophane group; more Si-rich species include the weeksite group, with a U:Si ratio of 2:5. The most U-rich species is soddyite with U:Si = 2:1 and the most Si-rich species is uranosilite with U:Si = 1:7.

Soddyite is one of only three uranyl silicates that do not contain other metal cations (the others being uranosilite and swamboite). It typically forms distinct dipyramidal crystals or elongated prisms. The crystals may form subparallel groups or divergent clusters; massive, fibrous, and earthy forms have also been reported. Opaque crystals are dull to bright yellow; transparent crystals are yellow to dark amber. In Congo, soddyite is common at Shinkolobwe (TL) as dipyramidal crystals up to 3 mm associated with sklodowskite, kasolite, curite, and metatorbernite; also in massive form associated with curite. At Musonoi and at Kolongwe it is associated with cuprosklodowskite and vandenbradeite; it is particularly abundant at swambo, associated with sklodowskite and black base-metal oxides. Other locales include: the Ruggles mine, Grafton Center, NH, as pseudomorphs after uraninite; the Lucky Mc mine, Fremont Co., WY; the Jackpile mine, Grants, NM; the Steel City mine, AZ; Honeycomb Hills, UT; and a pegmatite at Norabees, Namaqualand, South Africa.

## The Uranophane Group

Members of this group are hydrated uranyl silicates with the following general formula: $A[UO_2SiO_4]_y \cdot nH_2O$, where A = $Pb^{+2}$, $Ca^{+2}$, K, Na, $Cu^{+2}$, $Mg^{+2}$, $Co^{+2}$, $U^{+6}$. The group includes kasolite, uranophane, sklodowskite, cuprosklodowskite, boltwoodite, oursinite, swamboite, and uranophane-ß.

Fig. 160. Soddyite crystals from the Jackpile mine, Grants, New Mexico, showing interesting color zoning with opaque, bright yellow cores and a darker transparent outer zone. *RJL343*

Fig. 161. Small golden yellow soddyite crystals from Swambo, Congo. RJL901

Boltwoodite, the potassium analogue of uranophane, was the first reported alkali uranyl silicate. It forms pale yellow acicular crystals that may be mistaken for sklodowskite or uranophane. Individual crystals have a pearly luster on cleavage surfaces; radial aggregates may appear vitreous to silky depending on crystallite size. It typically forms coatings in fractures and cavities as well as compact to earthy pseudomorphs after uraninite. Depending on the space available for crystal growth, divergent tufts or sprays, flattened radial aggregates, or complete hemispherical masses may be found. Boltwoodite is fluorescent dull green under both LW and SW UV light. Boltwoodite is a fairly common secondary mineral that has been recognized at numerous localities in varied geological environments. At the Delta mine, Emery Co., UT (TL) it forms yellow wartlike aggregates of fibers coating sandstone as well as in other specimens associated with becquerelite, fourmarierite, and uranyl sulfates. In the Williams quarry, Easton, PA it forms as an alteration product of thorian uraninite in serpentine. Pseudomorphs after uraninite from Alto Boqueirao, Brazil, and from the Little Indian mine, CO are densely microcrystalline and pale straw yellow. Colorful micromount specimens from the New Method mine, near Amboy, CA, feature pale yellow boltwoodite associated with quartz and minute clusters of very dark purple fluorite crystals. Rössing, Namibia is the source of what are at present arguably the best specimens of boltwoodite, in radial aggregates up to 2 cm across associated with clear quartz, calcite, gypsum, and other minerals.

Fig. 162. A hand specimen of boltwoodite on sandstone from the type locale, the Delta mine, Utah. The boltwoodite has formed a thick, earthy yellow efflorescence on the rock. *RJL893*

Fig. 163. Bright yellow boltwoodite forming dense fans of acicular crystals on dark calcite from Rössing, Namibia. This habit is fairly common and contrasts somewhat with the divergent sprays of crystals seen in the specimen in Figure 43, also from the Rössing deposit. *RJL2590*

Fig. 164. Another habit of boltwoodite that is sometimes seen at Rössing, Namibia: a spherical aggregate of orange, nearly transparent crystals, also on the typical dark calcite lining a cavity in the country rock. *RJL2589*

Fig. 165. Minute, pale yellow acicular boltwoodite with very dark fluorite and quartz from the New Method mine, near Amboy, San Bernardino Co., California. *RJL798*

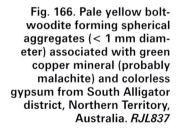

Fig. 166. Pale yellow boltwoodite forming spherical aggregates (< 1 mm diameter) associated with green copper mineral (probably malachite) and colorless gypsum from South Alligator district, Northern Territory, Australia. *RJL837*

Cuprosklodowskite forms acicular prismatic crystals elongated on the c-axis, in stellate fibrous aggregates of crystals up to 1 cm long, silky to matted crusts, and radial-fibrous aggregates and nodules. It is pale green to emerald green, also olive green; fibrous samples can be nearly white. Cuprosklodowskite is not fluorescent. It forms by the oxidation of deposits containing primary uraninite and copper minerals. It is found in most of the copper-cobalt seposits of southern Shaba, including Kolongwe (TL), Kambove, Kasompi, Kolwezi, Luiswishi, Musonoi, and Shinkolobwe. The most common associates are vandenbrandeite, kasolite, malachite, and sklodowskite; less frequently seen are rutherfordine, schoepite, and other secondary minerals. It is found at Grants, NM, as well as at several localities in Utah, including Red Canyon and White Canyon in San Juan Co. and at Seven Mile Canyon near Moab. Other localities include: the Nicholson mine, Lake Athabaska, SASK, Canada; Johanngeorgstadt, Germany; Jachymov, Czech Republic; and Amelal, Morocco.

Fig. 167. A spectacular example of cuprosklodowskite from Musonoi, Congo, showing radiating aggregates of very long, thin crystals on iron-stained copper-rich ore. *RJL2942*

Fig. 168. Bright apple green cuprosklodowskite needles thickly filling small seams and cavities from Musonoi, Congo. Dark green malachite and minor rutherfordine (yellow) are also present. *RJL348*

*110*

Kasolite forms prismatic crystals to 4 mm long, elongated and flattened lathlike crystals, flattened radial clusters, and earthy masses. It is one of the constituents of *gummite* as an alteration or pseudomorph after uraninite. It may be various shades of yellow to orange with a subadamantine luster; it is not fluorescent. It is very common at Kasolo, Congo (TL), most typically as divergent sprays or sheaflike aggregates associated with uranophane, curite, and various uranyl phosphates. Elsewhere in Congo, it is commonly seen at Menda, Swambo, Luiswishi, Kambove, Kalongwe, and Musonoi; it is also found at Franceville, Gabon. Other locales include: the Ruggles mine, NH, in gummite; in the East Walker River area, Lyon Co., and at Goodsprings, Clarke Co., NV; the Nicholson mine, SASK, and Great Bear Lake, NWT, Canada. It has been found at a number of localities in France, including Kersagalec, Ligud, Marbihon; Grury, Saone-et-Loire; and Puy de Dome.

Fig. 169. Orange-brown kasolite crystals in radiating aggregates forming a patch about 2 cm wide from Shinkolobwe, Congo. *RJL1371*

Fig. 170. Clusters of tabular orange kasolite with bright green torbernite from Musonoi, Congo. *RJL1026*

Fig. 171. Scanning electron micrograph of a small crystal group removed from the specimen in Figure 170, showing several kasolite crystals with a small torbernite crystal attached at upper left. *RJL1026*

Fig. 173. Detail of the specimen shown in Figure 172, showing prismatic orange kasolite crystals thickly lining the surface of a small cavity. *RJL332*

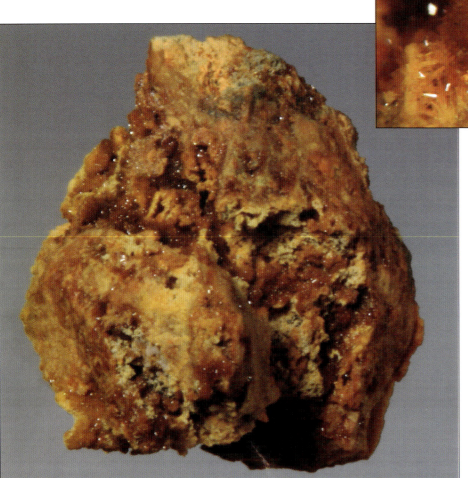

Fig. 172. Cavernous mass of nearly pure kasolite with drusy orange crystals lining cavities and fissures from Musonoi, Congo. *RJL332*

Sklodowskite forms acicular prismatic crystals that commonly have a tapering termination. It can form rosettes, radial fibrous aggregates, mammilary crusts, and granular to earthy masses. It is pale to lemon yellow; fibrous samples can be nearly white; it is not fluorescent. Sklodowskite occurs in the oxidation zone of uranium deposits, often replacing uraninite, and generally associated with other secondary uranium species. In the U.S., it occurs at the New Haven mine, Crook Co., WY; the Oyler Tunnel claim, near Fruita, and at Honeycomb Hills, Juab Co., UT; the Midnite mine, Stevens Co., WA; and in the Grants area, NM. It is found as bright yellow spherical tufts on transparent gypsum at Charcas,

SLP, and from Naica, CHIH, Mexico. In the Congo, at Shinkolobwe (TL) it forms bright yellow spherical nodules of minute fibers, radiating clusters of pale yellow acicular crystals, and massive yellowish-white replacements of uraninite cubes. Associates include hydrous uranium oxides such as becquerelite, billietite, schoepite, and curite, along with other uranyl silicates and rutherfordine. At Kalongwe, it occurs with secondary minerals of uranium and copper (cuprosklodowskite and vandenbrandeite). It is also found at Luiswishi, Musonoi, and Swambo. Fine specimens have been reported from the Pedra Preta magnesite mine, Brumado, BAH, Brazil in association with novačekite and zeunerite.

Fig. 174. Bright yellow sklodowskite needles forming spheres about 2 mm wide on colorless gypsum from Charcas, San Luis Potosi, Mexico. *RJL178*

Fig. 175. An extraordinary specimen observed by the author during examination of samples from Musonoi, Congo in collaboration with Sharon Cisneros of Mineralogical Research Co.: golden yellow elongated prismatic crystals of sklodowskite with minute colorless acicular crystals (apparently uranophane) on their tips from Musonoi, Congo. *RJL912*

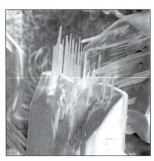

Fig. 176. Scanning electron micrograph of the material shown in Figure 175, showing minute lathlike crystals emerging from the terminations of sklodowskite crystals.

Uranophane forms acicular crystals elongated on [010], often terminated by an oblique face that forms angles of 95° to (100) and 93° to (001). The crystals may form radiating to spherical rosettes, felted masses and coatings, or comprise a massive yellow layer typically associated with orange-brown *gummite*. At Miedzianca (Kupferberg), Poland (TL), it was found both as dense pseudomorphs after uraninite and as tufted aggregates in the oxidized zone of the copper veins. Material from Wölsendorf, Bavaria was later described under the name *uranotile*; subsequent analyses showed that the two species were identical.

Fig. 177. Colorful specimen of bright yellow uranophane on pink cobaltian dolomite from Musonoi, Congo. *RJL1133*

Localities where uranophane forms alteration rims with *gummite* include: Grafton, NH; Avondale, PA; the Spruce Pine district, NC; and the Black Hills, SD. Many localities in the Grants district, NM, have produced excellent specimens; the occurrences in the Todilto limestone at Haystack Butte are noteworthy. At Wölsendorf, Bavaria, it is abundant as crusts of acicular crystals and as fibrous nodules in cavities in massive purple fluorite; elsewhere in Germany, it is found at Schneeberg and at Oberpfalz. In Canada, at the Faraday mine, Bancroft, ONT, silky crystals form radial clusters to several cm. Uranophane is widespread in the deposits of southern Shaba; it has been collected from Kalongwe, Kambove, Kolwezi, Luiswishi, Musonoi, Swambo, and Shinkolobwe. The felted masses can attain several cm and often serve as the base on which other secondary uranium species crystallize. It is preferentially associated with hydrous uranium oxides, kasolite, and rutherfordine adjacent to primary uraninite. At Swambo and Kolwezi it is often associated with soddyite. At Musonoi, colorful specimens of bright yellow uranophane on pink cobaltoan dolomite are found.

Fig. 178. Felted masses of yellow uranophane on quartz from the Hanosh mine, Grants, New Mexico. RJL333

Fig. 179. Bright yellow velvety uranophane forming a rich crust on massive black uraninite from Shinkolobwe, Congo. *RJL800*

Fig. 180. Dull yellow needles of uranophane to about 1 cm covering matrix from Milan, Valencia Co., New Mexico. *RJL740*

Uranophane-ß is a hydrous monoclinic calcium uranyl silicate dimorphous with uranophane. It has been found as velvety or bristly coatings, or radial to fanlike aggregates of acicular crystals. More compact or feltlike fiber aggregates are often seen as a replacement of uraninite. Individual needles are often somewhat thicker than those of uranophane or sklodowskite, elongated on the c-axis with a square or rectangular cross section, sometimes terminated with a relatively large {001} face. The three minerals cannot be reliably distinguished by appearance alone; either measurement of the optical constants or X-ray crystallography will provide a reliable identification. Uranophane-ß has the same occurrences and associations as uranophane, but is less common. Uranophane appears to be the more stable phase at ordinary temperature and pressure, based on the observation that when it is finely crushed uranophane-ß may revert to uranophane. The Faraday mine, near Bancroft, Ontario has produced many excellent specimens of radiating, canary-yellow needles in a matrix of brecciated and hematitized granite, associated with clear gypsum. Noteworthy American localities include: the Dunton mine, Newry, ME; the Ruggles mine, Grafton, NH; Haddam Neck, East Hampton, CT; Jim Thorpe (Mauch Chunk) PA; Crown Point Center, NY; the Grants District, NM; the Silver Star District, Mineral Co., NV; and the Cameron area, Coconino Co., AZ.

Fig. 181. Hand specimen about 10 cm wide with deep golden needles of uranophane-ß thickly covering the surface, from Rössing, Namibia. *RJL905*

At Jachymov, Czech Republic (TL) it forms crusts of tiny needles on altered uraninite in association with liebigite and calcite. In Germany it has been reported from: Menzenschwand, Black Forest; Wölsendorf, Bavaria; the Ellwieler, Katharina I, and Katharina II mines and the Schweisweiler-Winnweiler uranium deposit in Rhineland-Palatinate; and the Mansfeld Basin in Saxony-Anhalt. French localities include: the Bigay mine, Auvergne; Le Bauzot mine, Burgundy; Mas d'Alary and Rabejac in Lodeve; Margnac and Margnac II mines in Compreignac; and Les Bois-Noirs mine in Loire. In Africa, it is found intergrown with uranophane at Shinkolobwe, Congo and as fine yellow-brown crystals from Rössing, Namibia.

## The Weeksite Group

The weeksite group minerals have the general formula $M(UO_2)_2(Si_5O_{13})\cdot nH_2O$, where M = Ca, Mg, or K+Na, and n = 4-5.

Fig. 182. Another habit of uranophane-ß from Rössing, Namibia, showing divergent sprays of coarser, amber crystals filling a cavity in the country rock. *RJL2568*

Fig. 183. Detail of the specimen in Figure 182 from Rössing, Namibia. *RJL2568*

These minerals are relatively rare and form in arid environments, possibly from the evaporation of Si-rich waters of relatively high pH, in which dissolved polymeric silicate species could reach fairly high concentrations. In contrast to the uranyl sulfates and carbonates, the weeksite–group minerals have much less tendency to dissolve on re-exposure to fresh water. Like uranophane, upon exposure to carbonate-free water, the minerals might preferentially lose some Ca and Si, thereby altering to more U-rich species such as uranophane or soddyite. Conversely, in more alkaline, carbonate-rich waters, these minerals might preferentially lose U, altering to cryptocrystalline or amorphous silica (Finch and Murakami 1999). The group includes haiweeite, meta-haiweeite, Mg-haiweeite, and weeksite. The related mineral coutinhoite is a thorium uranyl silicate probably isostructural with weeksite.

Haiweeite forms very small spherulitic aggregates of minute acicular crystals or small flake-like grains. It is typically pale yellow to greenish yellow and is weakly fluorescent dull green under UV light. At the Haiwee Reservoir, Coso Mts., CA (TL) haiweeite occurs on fracture surfaces in granite and in voids in the nearby lake-bed sediments. The outward appearance of the material is typically earthy, although under the stereo microscope individual needles may often be seen. Far better examples of the species occur in Brazil: at Perus, north of São Paulo, it is found along fractures in a tourmaline granite associated with autunite, meta-autunite, uranophane, uranophane-ß, phosphuranylite, torbernite, meta-torbernite, and uranium opal. It also occurs in crevices in gneiss near Bad Gastein, Salzburg, Austria, and in the Hingyotoge deposit Tottori Pref., Japan. Meta-haiweeite is a species described mainly from studies of synthetic material in which haiweeite was artificially dehydrated in the laboratory by heating. Its properties are similar to those of haiweeite and the two phases are reported to be intimately associated at the type locality.

Fig. 184. Patches of pale yellow acicular to massive haiweeite on stained quartz from the Coso uranium mine, Inyo Co., California. *RJL695*

Fig. 185. Light yellow spherical aggregates of haiweeite scattered on matrix from Teofilo Otoni, Minas Gerais, Brazil. *RJL2719*

Weeksite typically forms small spherules of radiating lathlike crystals, which are yellow with a waxy to silky luster. At the type locality, weeksite occurs in opal veinlets in rhyolite with carbonates and gypsum. Weeksite has been reported from a number of American localities, including: the Autunite No. 8 claim, Thomas Mountains, UT (TL); the Mammoth mine, near Presidio, TX; the Jackpile mine, NM; the Anderson mine, Mojave Co., AZ; the Goodwill claim, UT; and at Haiwee Reservoir, CA. A small number of excellent specimens were found at an ore prospect near Rössing, Swakop River Valley, Namibia not far from the better-known uranium mine, which is the source of fine boltwoodite. It also occurs at Margaritas deposit, Sierra Peña Blanca, CHIH, Mexico; at Les Bois Noirs, France; Yinnietharra, WA, Australia; and localities in Japan, Afghanistan, and Russia.

Uranosilite is the most Si-rich uranyl silicate; it is very rare. This mineral occurs as minute crystals, acicular on [001], intimately intergrown with uranophane and studtite on quartz and hematite at Menzenschwand, Schwartzwald, Germany (Walenta, 1983).

Fig. 186. Golden yellow spherical aggregates of weeksite with pale yellow acicular boltwoodite from an ore prospect near Rössing, Namibia. *RJL2864*

Fig. 187. Another sample from the same locale as the specimen in Figure 186. In the microscope the weeksite and boltwoodite can be easily distinguished by differences in their color and habit. *RJL2704*

Fig. 188. Deep yellow acicular weeksite forming spherical aggregates on matrix from the Anderson mine, Arizona. *RJL3140*

# • Other Minerals Containing Essential U or Th

### The Monazite Group

The monazite group is a series of phosphates and arsenates with the general formula $ATO_4$ where $A = Ce^{3+}$, La, Nd, $Th^{4+}$, Ca, Bi, and REE; and $T = P$ or As. One member, cheralite, contains essential Th in its structure. Other members of the monazite group may contain significant amounts of Th as an impurity.

Cheralite forms crude translucent green crystals in masses up to 5 cm in a kaolinized pegmatite at Kultakughi, Travancore, India (TL). *Brabantite*, which had been regarded as a separate species, falls within the presently accepted definition of cheralite. It forms elongated gray-brown crystals, often with a lighter brown surface alteration. It has been reported from Brabanite farm, Karibib district, Namibia; and Xingjiang, China.

### The Rhabdophane Group

The rhabdophane group is a series of phosphate monohydrates with the general formula $APO_4 \cdot H_2O$ where $A = Ce$, La, Nd, Th, Ca, U, and $Fe^{3+}$. They may be thought of as the hydrated equivalents of the monazite group. Members of the group containing essential U or Th include: brockite, grayite, ningyoite, and tristramite. The minerals generally occur as earthy to cryptocrystalline coatings and none forms large crystals.

### The Pyrochlore Group

The pyrochlore group is a series of cubic oxides that contain essential amounts of niobium, tantalum, or titanium. Pyrochlore was first described in 1826, microlite in 1835, and "hatchettolite" in 1877. Over the next 100 years, many analyses were reported and because of the large number of compositional variations, the number of named species and varieties began to proliferate, along with several proposed schemes for classifying the group. The classification used here follows the recommended nomenclature for the pyrochlore group adopted by the IMA Commission on New Minerals and Mineral Names (Hogarth, 1977). The members of this group that contain essential uranium are: betafite, bismutopyrochlore, uranmicrolite, and uranopyrochlore. Of these, betafite is of most interest to collectors.

Betafite was first described by Lacroix in 1912, who named it for its type locale, Betafo, Madagascar (Frondel 1958). It typically forms sharp octahedral crystals, which are usually metamict. At the type locality, crystals as large as 15 cm have been found, although most crystals are commonly less than a few cm. Betafite is opaque, black to brown, and fresh crystals have a waxy or greasy luster; altered material often has a dull, yellowish surface layer that is somewhat hydrated and depleted in uranium (Frondel 1958). It is found as a primary mineral in pegmatites, associated with fergusonite, thorite, rutile, euxenite, cyrtolite, allanite, beryl, titanite, and magnetite. Excellent, large crystals are found with zircon at the Silver Crater Mine, Bancroft, Ontario. In the United States it occurs at the Brown Derby No. 1 mine, Gunnison, Colorado with gahnite, monazite, and columbite-tantalite species. Small crystals are found with uranoan zircon (*cyrtolite*) in the Cady Mountains near Hector, California; also in the Pidlite pegmatite, Mora County, New Mexico. Exceptional specimens have also been found at Mina El Muerto, Telixtlahuaca, Mexico.

**Fig. 189. Complex dodecahedral crystals of betafite on corroded hornblende from the Halo uranium mines, Ontario, Canada.** *RJL2151*

Fig. 190. Simple octahedral crystal of betafite with pale tan surface alteration from the type locale, Betafo, Madagascar. This specimen has an interesting history: In 1907 LaCroix collected a large lot of crystals, on which he based the definition of the species. The samples were taken to Denmark and later sent to England during World War II. Fifty years later the lot was auctioned and these historic samples became available to collectors. *RJL2564*

## The Samarskite Group

Samarskite-group minerals are ordered, mixed-oxide compounds having the general formula $A^{+3}B^{+5}O_4$, and a structure that is related to that of $\alpha$-$PbO_2$. In this formula, A = Ca, U, Th, Y, REE, and Fe; and B = Nb, Ta, and Ti. The compositional variations within this group were recently reviewed (Hanson et al. 1999) and based on that analysis, it was proposed that the group includes three possible species based on the occupancies of the A site. If REE+Y is dominant, the mineral should be named samarskite-(REE+Y), where the actual suffix used will be the dominant rare earth cation, in accordance with Levinson notation. If U+Th is dominant, the species is named ishikawaite. If Ca is dominant, the mineral is named calciosamarskite. In all cases Nb > Ta and Ti on the B site.

Analysis of these minerals can be difficult because they are usually metamict and may be altered on the surface. The crystal structure can be restored by annealing, but it is not always certain that the restored sample is identical in all respects to the mineral that crystallized originally. None can be identified by visual inspection or through simple field tests. Short of actual microprobe analysis of the sample in question, the collector must often rely on whatever published information exists on the specific locality and make an educated guess.

Calciosamarskite occurs as a minor accessory mineral in pegmatites. It typically forms black glassy anhedral masses with a vitreous luster and a brown to black streak. The mineral is metamict and shows a conchoidal fracture. It was found at Wiseman's mica mine in Mitchell Co., NC, associated with columbite, uranmicrolite, and muscovite; it has also been reported from Spruce Pine, and other localities in NC.

Ishikawaite was named for the locality, Ishikawa District, Fukushima, Iwaki Pref., Japan, where it occurs in pegmatite. The mineral forms black, glassy, anhedral masses with a vitreous luster. The mineral is partially or completely metamict and has a conchoidal fracture. Recently, samples resembling the classic description of "samarskite" were collected from various localities in the Topsham Pegmatite District, Maine; these sites included the Yedlin Location, the Russell Brothers Quarry, Stand Pipe Hill, and the Swamp No. 1 Quarry. In all cases, analysis showed U+Th greater than REE+Y so these materials are all properly regarded as ishikawaite (Hanson et al., 1998).

Samarskite-(Y) forms rough tabular or prismatic crystals or masses in granite pegmatites. It is black or very dark brown, often with a yellowish-brown surface alteration. Some notable localiies include: Miask, Ural Mts., Russia (TL); Antanamalaza, Madagascar; Divina de Uba, MG, Brazil; and Spruce Pine, NC. A "uranium-rich samarskite" from Kunar, Afghanistan is very similar to the material from Ishikawa, but its composition places it within the range of samarskite-(Y) as presently defined.

Some other mixed-oxide minerals that are worthy of note include brannerite, orthobrannerite, thorutite, and davidite. Several are important ore minerals and some occasionally produce large crystals of interest to collectors.

Brannerite is the third most important uranium ore mineral (after uraninite and coffinite). It forms indistinct prismatic crystals or rounded grains that are black when fresh; surface alterations may be olive-green, brown, or yellow. It is usually metamict. Brannerite is a common accessory mineral in uraninite and coffinite ore deposits. Notable occurrences include: as alluvial grains at Kelly Gulch, Custer Co., ID (TL); in quartz pebble conglomerate in the Blind River district, Canada; as very large crystals in pegmatite at Fuenteovejuna, Cordoba, Spain; in quartz veins in granitoid rock at Crocker's Well, SA, Australia; and in Witwatersrand, South Africa.

Thorutite is the thorium analogue of brannerite. It forms short black crystals that are generally metamict. At the Kutyur-Tyube thorium deposit, Alai Range, southern Kyrgyzstan (TL) crystals to 2 cm are associated with thorite, zircon, calcite, barite, and galena in microcline-nepheline veins in syenite (Pekov 1998). It has also been reported from the magnesite deposit at Brumado, BAH, Brazil.

Fig. 191. A tabular crystal of ishikawaite about 6 mm long in pegmatite from Yedlin quarry, Topsham, Maine. *RJL1792*

Davidite is a rare earth titanium iron oxide containing uranium, which has been mined as a uranium ore at several localities. The La-dominant species, davidite-(La), was mined at Radium Hill, SA, Australia (TL) where it formed rough crystals and irregular masses in high temperature hydrothermal lodes, associated with quartz, hematite, ilmenite, rutile, magnetite, and biotite. Large crystals that assayed up to 8% $U_3O_8$ were mined at Mavuzi, near Tete, Mozambique; numerous smaller deposits have been identified within the Mavuzi district (Heinrich 1958). Other locales include: Bek-Tau Ata, Kazakhstan; the Faraday mine, Bancroft, ONT, Canada; and the Pandora prospect, Pima Co., AZ. The Ce-dominant species, davidite-(Ce) is found at Tuftane, Norway (TL) and at the Faraday mine and the Foster Lakes area, Canada.

Other minerals containing essential U or Th are listed in the appendix.

Fig. 192. A group of davidite-(La) crystals showing tan surface alteration from Bektau-Ata, Kazakhstan. *RJL2801*

# • Minerals Frequently Containing U or Th as Impurities

Allanite refers to several members of the epidote group. Under modern nomenclature, the type material from Greenland is allanite-(Ce). Later workers described additional compositions that are now denoted allanite-(La) and allanite-(Y). Allanites are a common accessory mineral in granite, granodiorite, monzonite, syenite, and granitic pegmatite. They are also found in diorite and gabbro, as phenocrysts in acid volcanic rocks, and in a number of metamorphic rock types. Many allanites contain uranium and thorium in addition to rare earth elements. According to the literature reviewed by Gieré and Sorensen (2004) the maximum reported $ThO_2$ content is 4.9 wt.% which would correspond to about 0.07 Th atoms per formula unit (apfu). The maximum $UO_2$ content is 0.82 wt.% in a crystal that also contained 1.09 wt.% $ThO_2$, corresponding to 0.02 apfu of each.

Many REE-bearing epidote minerals are partly or completely metamict. One traditional approach to studying metamict minerals is to anneal the sample in order to recrystallize or "heal" the radiation damage and thereby restore the crystal structure for X-ray analysis. The problem is that a metamict mineral is fairly reactive and may adsorb soluble elements from the environment or undergo ion exchange. Thus, the "restored" crystal might have a composition that differs significantly from that of the original crystal and in effect becomes a synthetic material. As stated by Armbruster et al. (2006): "There is at least some suspicion that such 'mineral' compositions are influenced by the experimenter and are not an unaltered product of nature. These problems are not specific of epidote-group minerals but are much more prominent in other mineral groups with higher concentrations of radioisotopes. ... We recommend exercising caution with compositions of 'partly' metamict epidote-group minerals in naming new species, even if the 'faulty' lattice has been mended by subsequent heat treatment."

Baddeleyite, $ZrO_2$, an important ore of zirconium, reportedly "may contain appreciable thorium and small amounts of uranium" (Nininger 1954).

The rare earth carbonates ancylite-(Ce), bastnaesite-(Ce), cordylite-(Ce), synchysite-(Ce), lanthanite-(La), parisite-(Ce), and tengerite-(Y) may contain from 0.2 to 1.6% $ThO_2$ and lesser amounts of U. The phosphate mineral xenotime, $YPO_4$, may contain ~5% $UO_2$ and ~3% $ThO_2$ (Heinrich 1958).

Radioactive barites, thought to contain Ra, Th, and U, are found at several hot spring deposits, including: Hokuto, Taiwan; Shibukuro, Japan; and Teplitz, Czech Republic. *Radiobarite* has also been noted in the upper parts of the Tyuya Muyun deposit, Kyrgyzstan, along with "radiocarbonate" of presumed composition $(Ca,Ra)CO_3$ (Heinrich 1958).

Zircon is a common accessory mineral in granitic and syenitic igneous rocks. Because of its hardness and chemical stability it is also a common detrital mineral. In granitic rocks, U and Th are often concentrated in the zircon and some other accessory minerals. Uranoan and thorian varieties of zircon have been given names such as *cyrtolite*, *malacon*, *orvillite*, etc. The radioactive varieties are often partly or completely metamict; they tend to have lower density and are more easily altered by corrosive solutions. Fine crystals of zircon (var. *cyrtolite*) found at several sites near Bancroft, ONT, Canada typically contain 1-2% $UO_2$ and 4-5% $H_2O$. They are often lighter colored than other zircons and typically have curved faces. Well-formed crystals are found with fergusonite at the John Gole quarry. Doubly terminated crystals to 1.5 cm are found along with betafite in black humus in the Silver Crater Mine. Fine large crystal clusters have been collected from the Davis nepheline quarry in Monteagle Township (Kennedy, 1979).

Fig. 193 A large group of curved brown crystals of zircon var. cyrtolite from the Davis mine, Bancroft, Ontario, Canada. *RJL2660*

# Bibliography

## • Suggested Core Library for Further Reading

Beginning around 1948 the U.S. Atomic Energy Commission supported a rigorous study of uranium and thorium minerals as part of a broader effort to understand the nature and geology of the raw materials for nuclear power programs. Among several noteworthy reports published by the USGS (Weeks and Thompson 1954; C. Frondel 1958; J. Frondel et al. 1967), the seminal work of Clifford Frondel, *Systematic Mineralogy of Uranium and Thorium*, has rightly been described by Dr. Carl Francis of Harvard Museum as "the basis of our modern understanding of radioactive minerals." Another excellent volume from this period is *Mineralogy and Geology of Radioactive Raw Materials* (Heinrich 1958), which discusses the geology of several thousand deposits, organized by the type of formation.

The importance of uranium resources in Colonial Africa led to another series of books, *The Secondary Uranium Minerals of Zaire*, that are highly recommended for any serious collector or student of U minerals. Although published in French, the mineralogical data and photomicrographs are easily comprehensible to English-speaking readers. The Congo (Zaire) has been the source of many superb uranium minerals and is well known in the collecting community, making these volumes an indispensable resource. The original volume (Guillemin et al. 1958) contains detailed descriptions of the important minerals, along with 27 color plates, and is occasionally available from rare book dealers. A completely updated version, along with two supplements, (Deliens et al. 1981, Deliens et al. 1984, and Deliens et al. 1990), describes 66 minerals and their associations.

An up-to-date review of the scientific literature can be found in *Uranium: Mineralogy, Geochemistry, and the Environment*, which was recently published by the Mineralogical Society of America as Reviews in Mineralogy Vol. 38 (Burns and Finch 1999). This volume reviews such topics as crystal chemistry, systematics, paragenesis of deposits, analytical methods, and implications for technological problems such as contamination and the disposal of spent reactor fuel. The individual review papers provide further references to literally thousands of additional research reports, including many original descriptions of new U minerals.

## • References Cited

Armbruster, T., Bonazzi, P., Akasaka, M., Bermanec, V., Chopin, C., Giere, R., Heuss-Assbichler, S., Liebscher, A., Menchetti, S., Pan, Y., and Pasero, M. (2006) Recommended nomenclature of epidote-group minerals, *European Journal of Mineralogy*, 18, 551-67.

Berning, J., Cooke, R., Hiemstra, S., and Hoffman, U. (1976) The Rössing uranium deposit, South West Africa, *Economic Geology* 71, 351-368.

Birch, W., Mumme, W., and Segnit, E. (1988) Ulrichite – a new copper calcium uranium phosphate from Lake Boga, Victoria, Australia, *Australian Mineral.* 3,125-34; [see abstract (1990) *Amer. Mineral.* 75, 243].

Brugger, J., Ansermet, S., and Pring, A. (2003) Uranium minerals from Mt. Painter, Northern Flinders Ranges, South Australia, *Australian J. Mineral.* 9 [1], 15-31.

Brugger, J., Burns, P., and Meissner, N. (2003) Contribution to the mineralogy of acid drainage of Uranium minerals: Marecottite and the zippeite-group, *Amer. Mineral.* 88, 676-85.

Burns, P. and Finch, R. (1999) *Uranium: Mineralogy, Geochemistry, and the Environment*, *Rev. in Mineral*. 38, Mineralogical Society of America, Washington, DC.

Burns, P. and Finch, R. (1999a) Wyartite: crystallographic evidence for the first pentavalent-uranium mineral, *Amer. Mineral.* 84, 1456-60.

Burns, P. and Hanchar, J. (1999) The structure of masuyite, $Pb[(UO_2)_3O_3(OH)_2](H_2O)_3$, and its relationship to protasite, *Can. Mineral.* 37,1483-91.

Burns, P. and Hughes, K.-A. (2003) Studtite, $[(UO_2)(O_2)(H_2O)_2](H_2O)_2$: The first structure of a peroxide mineral, *Amer. Mineral.* 88, 1165-8.

Cesbron, F, and Bariand, P. (1975) The uranium-vanadium deposit of Mounana, Gabon, *Mineral. Record* 6 [5], 237-49.

Deliens, M., Piret, P., and Comblain, G. (1981) *Les Mineraux Secondaires D'Uranium Du Zaire*, Tervuren, Belgium.

Deliens, M., Piret, P., and Comblain, G. (1984) *Les Mineraux Secondaires D'Uranium Du Zaire (Supplement)*, Tervuren Belgium.

Deliens, M., Piret, P., and Van Der Meersche, E. (1990) *Les Mineraux Secondaires D'Uranium Du Zaire (2nd Supplement)*, Tervuren, Belgium.

DeVoto, R. (1978) Uranium in Phanerozoic sandstone and volcanic rocks, pp. 293-306, Min. Assoc. of Canada, *Short Course in Uranium Deposits: Their Mineralogy and Origin*, M. M. Kimberley, Ed., Univ. of Toronto Press.

Edwards, R. and Atkinson, K (1986) *Ore Deposit Geology and its Influence on Mineral Exploration*, 466 pp., Chapman and Hall, London.

Embrey, P. and Symes, R. (1987) *Minerals of Cornwall and Devon*, British Museum (Natural History), London.

Finch, W., Molina, P., Naumov, S., Ruzica, V., Barthel, F., Thoste, V., Muller-Kahle, E., Pecnik, M., and Thauchid, M. (1995) World Distribution of Uranium Deposits, First Edition; International Atomic Energy Agency, scale 1:30,000,000.

Finch, R. and Murakami, T. (1999) Systematics and paragenesis of uranium minerals, *Rev. in Mineral*. 38, 91-179, Min. Soc. Amer., Washington, DC.

Frondel, C. (1958) *Systematic Mineralogy of Uranium and Thorium* Geological Survey Bulletin 1064, U. S. Government Printing Office.

Frondel, C., Ito, J., Honea, R., and Weeks, A. (1976) Mineralogy of the zippeite group, *Can. Mineral.* 14, 429-36.

Frondel, J., Fleischer, M., and Jones, R. (1967) *Glossary of Uranium- and Thorium-Bearing Minerals, 4th Edition*, Geological Survey Bulletin 1250, U.S. Government Printing Office.

Gauthier-Lafay, F., and Weber, F. (1989) The Francevillian (Lower Proterozoic) uranium deposit of Gabon, *Econ. Geol*. 84, 2267-85.

George-Aniel, B., Leroy, J., and Poty, B. (1991) Volcanogenic uranium mineralizations in the Sierra Peña Blanca district0, Chihuahua, Mexico: three genetic models, *Econ. Geol.* 86, 233-48.

Giere, R., and Sorensen, S. 2004. Allanite and other REE-rich epidote-group minerals, *Reviews in Mineralogy and Geochemistry* 46: 431-93.

Guillemin, C., Destas, A., and Vaes, J. (1958) *Mineraux D'Uranium Du Haut Katanga*, Tervuren, Belgium.

Heinrich, E. (1958) *Mineralogy and Geology of Radioactive Raw Materials*, McGraw-Hill, New York.

Henry, D., Pogson, R., and Williams, P (2005) Threadgoldite from the South Alligator Valley uranium field, Northern Territory, Australia: second world occurrence, *Austral. J. Mineral*. 11 [1], 7-11.

Hoeve, J. and Sibbald, T. (1978) Mineralogy and geological settings of unconformity-type uranium deposits in northern Saskatchewan, pp. 457-74, Min. Assoc. of Canada, *Short Course in Uranium Deposits: Their Mineralogy and Origin*, M. M. Kimberley, Ed., Univ. of Toronto Press.

Hogarth, D. (1977) Classification and nomenclature of the pyrochlore group, *Amer. Mineral*. 62, 403-10.

Kennedy, I., (1979) Some interesting radioactive minerals from the Bancroft Area, Ontario. *Mineral. Rec*. 10 [3], 153-8.

Klepper, M. and Wyant, G. (1956) Uranium provinces, U. S. Geol. Survey Prof. Paper 300, 17-25.

Langford, F. (1978) Uranium deposits in Australia, pp. 205-16, Mineral. Assoc. of Canada, *Short Course in Uranium Deposits: Their Mineralogy and Origin*, M. M. Kimberley, Ed., Univ. of Toronto Press.

Langmuir, D. (1978) Uranium solution-mineral equilibria at low temperatures with applications to sedimentary ore deposits, pp. 17-55, Mineral. Assoc. of Canada, *Short Course in Uranium Deposits: Their Mineralogy and Origin*, M. M. Kimberley, Ed., Univ. of Toronto Press.

Lauf, R., Lindemer, T., and Pearson, R. (1984) Out-of-reactor studies of fission product – silicon carbide interactions in HTGR fuel particles, *J. Nucl. Materials* 120 [1], 6-30.

McMillan, R. (1978) Genetic aspects and classification of important Canadian uranium deposits, pp. 187-204, Min. Assoc. of Canada, *Short Course in Uranium Deposits: Their Mineralogy and Origin*, M. M. Kimberley, Ed., Univ. of Toronto Press.

Nash, J., Granger, H., and Adams, S. (1981) Geology and concepts of genesis of important types of uranium deposits, *Econ. Geol. 75th Anniv. Vol*., 63-116.

Nininger, R. (1954) *Minerals for Atomic Energy*, Van Nostrand, New York, 367 pp.

OECD (2004) *Uranium 2003: Resources, Production and Demand*, Organisation for Economic Co-operation and Development, Paris.

OECD (2006) *Uranium 2005: Resources, Production and Demand*, Organisation for Economic Co-operation and Development, Paris.

Ondrus, P., Veselovsky, F., Gabasova, A., Hlousek, J., and Srein, V. (2003a) Supplement to secondary and rock-forming minerals of the Jachymov ore district, *J. Czech Geological Soc*. 48 [3-4], 149-55.

Pabst, A., and Hutton, C. (1951) Huttonite, a new monoclinic thorium silicate, with an account of its occurrence, analysis, and properties, *Amer. Mineral*. 36, 60-69.

Pekov, I. (1998) *Minerals First Discovered on the Territory of the Former Soviet Union*, Ocean Pictures Ltd., Moscow, 369 pp.

Pekov, I. (2007) New minerals from former Soviet Union countries, *Mineralogical Almanac,* 11.

Robbins, M. (1994) *Fluorescence: Gems and Minerals under Fluorescent Light*, Geoscience Press, Phoenix, AZ, 374 pp.

Suzuki, Y., and Banfield, J. (1999) Geomicrobiology of uranium, *Rev. in Mineral*. 38, 393-432, Min. Soc. Amer., Washington, DC.

Threadgold, I. (1960) The mineral composition of some uranium ores from the South Alligator River area, Northern Territory, *Mineragraphic Investigations Technical Paper 2*, 53 pp., Commonwealth Scientific and Industrial Research Organization, Australia.

Vaes, J. (1947) Six nouveaux minereaux d'urane provenant de Shinkolobwe (Katanga) *Ann. Soc. Geol. Belgique* 70, B212-29 [see abstract (1948) *Amer. Mineral*. 33, 384].

Walenta, K. (1983) Uranosilite, a new mineral from the uranium deposit at Menzenschwand (Southern Black Forest) *Neues Jahr. fur Mineral., Monats*. 1983, 259-69; [see abstract (1984) *Amer. Mineral.* 69, 408-9].

Weeks, A. and Thompson, M. (1954) *Identification and Occurrence of Uranium and Vanadium Minerals From the Colorado Plateau*, U.S. Geol. Survey Bull. 1009-B.

Wenrich, K, Modreski, P., Zielinski, R., and Seeley, J. (1982) Margaritasite: a new mineral of hydrothermal origin from the Peña Blanca Uranium District, Mexico, *Amer. Mineral*. 67, 1273-89.

Zwaan, P., Arps, C., and de Grave, E. (1989) Vochtenite $(Fe^{2+},Mg)Fe^{3+}(UO_2)_4(PO_4)_4(OH) \cdot 12\text{-}13H_2O$, a new uranyl phosphate mineral from Wheal Basset, Redruth, Cornwall, England, *Mineral. Mag*. 53, 473-78; [see abstract (1990) *Amer. Mineral*. 75, 1212].

# Appendix:
# Checklist of Radioactive Minerals

| Mineral | Formula | Color | UV Fluorescence[a] |
|---|---|---|---|
| **Primary Oxides** | | | |
| Uraninite | $UO_2$ | black | |
| Thorianite | $ThO_2$ | black | |
| **Primary Silicates** | | | |
| Thorite | $(Th,U)SiO_4$ | black to brown | |
| Huttonite | $ThSiO_4$ | colorless to yellow-brown | pinkish white (SW) |
| Coffinite | $U(SiO_4)_{1-x}(OH)_{4x}$ | black | |
| Thorogummite | $Th(SiO_4)_{1-x}(OH)_{4x}$ | black | |
| **Secondary Hydrated Oxides** | | | |
| Agrinierite | $(K_2,Ca,Sr)U_3O_{10}\cdot4H_2O$ | orange | |
| Bauranoite | $BaU_2O_7\cdot4\text{-}5H_2O$ | red-brown | |
| Becquerelite | $Ca(UO_2)_6O_4(OH)_6\cdot8H_2O$ | lemon yellow to orange-yellow | |
| Billietite | $Ba(UO_2)_6O_4(OH)_6\cdot8H_2O$ | yellow to yellow-brown | |
| Calciouranoite | $(Ca,Ba,Pb)U_2O_7\cdot5H_2O$ | brown to brown-orange | |
| Compreignacite | $K_2(UO_2)_6O_4(OH)_6\cdot7H_2O$ | yellow | |
| Curite | $Pb_2U_5O_{17}\cdot4H_2O$ | deep orange to red-orange | |
| Fourmarierite | $PbU_4O_{13}\cdot4H_2O$ | red-orange to deep red | |
| Ianthinite | $U_2(UO_2)_4O_6(OH)_4\cdot9H_2O$ | purple-brown to black | |
| Masuyite | $Pb(UO_2)_3O_3(OH)_2\cdot3H_2O$ | reddish orange to carmine red | |
| Metacalciouranoite | $(Ca,Na,Ba)U_2O_7\cdot2H_2O$ | orange | |
| Metaschoepite | $(UO_2)_8O_2(OH)_{12}\cdot10H_2O$ | yellow | weak yellow-green |
| Metastudtite | $UO_2(OH)_4$ | pale yellow | |
| Protasite | $Ba(UO_2)_3O_3(OH)_2\cdot3H_2O$ | orange | |
| Rameauite | $KCaU_6O_{20}\cdot9H_2O$ | orange | |
| Richetite | $Pb_9(UO_2)_{36}(OH)_{24}O_{36}$ | black to deep brown | |
| Schoepite | $(UO_2)_8O_2(OH)_{12}\cdot12H_2O$ | sulfur yellow to amber | yellow-green |
| Spriggite | $Pb_3(UO_2)_6O_8(OH)_2\cdot3H_2O$ | orange | |
| Studtite | $UO_4\cdot4H_2O$ | pale yellow to cream | moderate yellow-green (SW) |
| Uranosphaerite | $BiO(UO_2)(OH)_3$ | yellow-orange to reddish orange | |
| Vandenbrandeite | $Cu(UO_2)(OH)_4$ | dark green | |
| Vandendriesscheite | $Pb_{1.5}(UO_2)_{10}O_6(OH)_{11}\cdot11H_2O$ | orange to amber | |
| Wölsendorfite | $Pb_7(UO_2)_{14}O_{19}(OH)_4\cdot12H_2O$ | orange-red to carmine | |
| **Secondary Carbonates** | | | |
| Albrechtschraufite | $Ca_4Mg(UO_2)(CO_3)_6F_2\cdot17H_2O$ | yellow-green | |
| Andersonite | $Na_2Ca(UO_2)(CO_3)_3\cdot6H_2O$ | yellow-green | yellow-green (SW,LW) |
| Bayleyite | $Mg_2(UO_2)(CO_3)_3\cdot18H_2O$ | yellow | weak yellowish green |

| Mineral | Formula | Color | UV Fluorescence[a] |
|---|---|---|---|
| Bijvoetite-(Y) | $(Y,Dy)_2(UO_2)_4(CO_3)_4(OH)_6 \cdot 11H_2O$ | yellow | |
| Blatonite | $UO_2CO_3H_2O$ | canary yellow | strong green-yellow (SW) |
| Čejkaite | $Na_4(UO_2)(CO_3)_3$ | pale yellow to beige | yellow to yellow-green |
| Fontanite | $Ca(UO_2)_3(CO_3)_4 \cdot 3H_2O$ | | deep yellow pale green (LW) |
| Joliotite | $(UO_2)(CO_3) \cdot nH_2O$ | yellow | |
| Kamotoite-(Y) | $Y_2U_4(CO_3)_3O_{12} \cdot 14.5H_2O$ | bright yellow | green (SW,LW) |
| Lepersonnite-(Gd) | $Ca(Gd,Dy)_2(UO_2)_{24}O_{12}(CO_3)_8(SiO_4)_4 \cdot 60H_2O$ | bright yellow | |
| Liebigite | $Ca_2(UO_2)(CO_3)_3 \cdot 11H_2O$ | green to yellowish green | green (SW,LW) |
| Metazellerite | $Ca(UO_2)(CO_3)_2 \cdot 3H_2O$ | chalky yellow | |
| Rabbittite | $Ca_3Mg_3(UO_2)_2(CO_3)_6(OH)_4 \cdot 18H_2O$ | pale green | creamy yellow (SW) |
| Rutherfordine | $UO_2(CO_3)$ | pale yellow, yellow-brown to orange | |
| Schroeckingerite | $NaCa_3(UO_2)(CO_3)_3(SO_4)F10H_2O$ | greenish yellow | bright yellow-green (SW,LW) |
| Sharpite | $Ca(UO_2)_6(CO_3)_5(OH)_4 \cdot 6H_2O$ | greenish yellow to olive green | |
| Swartzite | $CaMg(UO_2)(CO_3)_3 \cdot 12H_2O$ | bright green | bright yellow-green |
| Voglite | $Ca_2Cu(UO_2)(CO_3)_4 \cdot 6H_2O$ | grass green to emerald green | |
| Widenmannite | $Pb_2(UO_2)(CO_3)_3$ | yellow | |
| Wyartite | $CaU^{5+}(UO_2)_2(CO_3)O_4(OH) \cdot 7H_2O$ | purplish black | |
| Zellerite | $Ca(UO_2)(CO_3)_2 \cdot 5H_2O$ | light lemon yellow | weak green (SW,LW) |
| Znucalite | $CaZn_{11}(UO_2)(CO_3)_3(OH)_{20} \cdot 4H_2O$ | white to pale yellow | yellow-green |

**Secondary Sulfates**

**Zippeite Group**

| Mineral | Formula | Color | UV Fluorescence[a] |
|---|---|---|---|
| Zippeite | $K(UO_2)_2(SO_4)(OH)_3H_2O$ | bright yellow to orange-yellow | green |
| Cobalt-zippeite | $Co_2(UO_2)_6(SO_4)_3(OH)_{10} \cdot 16H_2O$ | orange-yellow to rust | |
| Magnesium-zippeite | $Mg_2(UO_2)_6(SO_4)_3(OH)_{10} \cdot 16H_2O$ | yellow | |
| Nickel-zippeite | $Ni_2(UO_2)_6(SO_4)_3(OH)_{10} \cdot 16H_2O$ | yellow-orange to brown-yellow | |
| Sodium-zippeite | $Na_4(UO_2)_6(SO_4)_3(OH)_{10} \cdot 4H_2O$ | yellow | green |
| Zinc-zippeite | $Zn_2(UO_2)_6(SO_4)_3(OH)_{10} \cdot 16H_2O$ | yellow, orange, red-brown | |
| Marecottite | $Mg_3(H_2O)_{18}[(UO_2)_4O_3(OH)(SO_4)_2](H_2O)_{10}$ | yellow-orange | |
| Pseudojohannite | $Cu_{6.5}[(UO_2)_2O_4(SO_4)_2]_2(OH)_5 \cdot 25H_2O$ | olive green | |

**Other Sulfates**

| Mineral | Formula | Color | UV Fluorescence[a] |
|---|---|---|---|
| Deliensite | $Fe^{2+}(UO_2)_2(SO_4)_2(OH)_2 \cdot 3H_2O$ | pale yellow | |
| Jachymovite | $(UO_2)_8(SO_4)(OH)_{14} \cdot 13H_2O$ | yellow | |
| Johannite | $Cu(UO_2)_2(SO_4)_2(OH)_2 \cdot 8H_2O$ | green | |
| Rabejacite | $Ca(UO_2)_4(SO_4)_2(OH)_6 \cdot 6H_2O$ | yellow | light yellow |
| Uranopilite | $(UO_2)_6(SO_4)(OH)_{10} \cdot 12H_2O$ | yellow | strong yellow-green |
| Meta-uranopilite | $(UO_2)_6(SO_4)(OH)_{10} \cdot 5H_2O$ | yellow | yellowish green |

**Secondary Selenites**

| Mineral | Formula | Color | UV Fluorescence[a] |
|---|---|---|---|
| Demesmaekerite | $Pb_2Cu_5(UO_2)_2(SeO_3)_6(OH)_6 \cdot 2H_2O$ | bottle green | |
| Derriksite | $Cu_4(UO_2)(SeO_3)_2(OH)_6$ | green to bottle green | |
| Guilleminite | $Ba(UO_2)_3(SeO_3)_2O_2(H_2O)_3$ | canary yellow | |
| Haynesite | $(UO_2)_3(SeO_3)_2(OH)_2 \cdot 5H_2O$ | yellow | bright green (SW,LW) |
| Larisaite | $Na(H_3O)(UO_2)_3(SeO_3)_2O_2 \cdot 4H_2O$ | yellow | green (SW) |
| Marthozite | $Cu(UO_2)_3(SeO_3)_3(OH)_2 \cdot 7H_2O$ | yellowish green to greenish brown | |
| Piretite | $Ca(UO_2)_3(SeO_3)_2(OH)_4 \cdot 4H_2O$ | lemon yellow | |

**Secondary Tellurites**

| Mineral | Formula | Color | UV Fluorescence[a] |
|---|---|---|---|
| Cliffordite | $UTe_3O_9$ | sulfur yellow | |
| Schmitterite | $(UO_2)TeO_3$ | pale yellow | |
| Moctezumite | $Pb(UO_2)(TeO_3)_2$ | bright orange to orange-brown | |

**Secondary Arsenites**

| Mineral | Formula | Color | UV Fluorescence[a] |
|---|---|---|---|
| Chadwickite | $H(UO_2)(AsO_3)$ | yellow | |

| Mineral | Formula | Color | UV Fluorescence[a] |
|---|---|---|---|
| **Secondary Molybdates** | | | |
| Calcurmolite | $Ca(UO_2)_{3-4}(MoO_4)_3(OH)_{2-5}\cdot7\text{-}12H_2O$ | honey yellow | strong yellowish green |
| Cousinite | $Mg(UO_2)_2(MoO_4)(OH)_2\cdot5H_2O$ | black | |
| Deloryite | $Cu_4(UO_2)(MoO_4)_2(OH)_6$ | dark green to black | |
| Iriginite | $(UO_2)(Mo_2O_7)\cdot3H_2O$ | canary yellow | |
| Moluranite | $H_4U^{4+}(UO_2)_3(MoO_4)_7\cdot18H_2O$ | black | |
| Tengchongite | $Ca(UO_2)_6(MoO_4)_2O_5\cdot12H_2O$ | yellow | |
| Umohoite | $[(UO_2)MoO_4(H_2O)](H_2O)$ | black to bluish black | |
| **Secondary Tungstates** | | | |
| Uranotungstite | $(Fe^{2+},Ba,Pb)(UO_2)(WO_4)(OH)_4\cdot12H_2O$ | yellow to brown | |
| **Secondary Phosphates and Arsenates** | | | |
| **Autunite Group** | | | |
| Abernathyite | $K_2(UO_2)_2(AsO_4)_2\cdot8H_2O$ | yellow | moderate yellow-green (SW,LW) |
| Autunite | $Ca(UO_2)_2(PO_4)_2\cdot10\text{-}12H_2O$ | yellow to green | bright yellowish green (SW,LW) |
| Bassetite | $Fe^{2+}(UO_2)_2(PO_4)_2\cdot8H_2O$ | green to brown | |
| Chernikovite | $(H_3O)_2(UO_2)_2(PO_4)_2\cdot6H_2O$ | yellow | bright yellowish green (SW,LW) |
| Heinrichite | $Ba(UO_2)_2(AsO_4)_2\cdot10\text{-}12H_2O$ | yellow to yellow-green | bright yellowish green (SW,LW) |
| Kahlerite | $Fe^{2+}(UO_2)_2(AsO_4)_2\cdot10\text{-}12H_2O$ | lemon yellow to yellow-green | |
| Lehnerite | $Mn^{2+}(UO_2)_2(PO_4)_2\cdot8H_2O$ | bronze-yellow | |
| Meta-ankoleite | $K_2(UO_2)_2(PO_4)_2\cdot6H_2O$ | yellow | yellowish green (SW,LW) |
| Meta-autunite | $Ca(UO_2)_2(PO_4)_2\cdot2\text{-}6H_2O$ | yellow | bright yellowish green (SW,LW) |
| Metaheinrichite | $Ba(UO_2)_2(AsO_4)_2\cdot8H_2O$ | yellow to green | bright yellowish green (SW,LW) |
| Metakahlerite | $Fe^{2+}(UO_2)_2(AsO_4)_2\cdot8H_2O$ | sulfur yellow | |
| Metakirchheimerite | $Co(UO_2)_2(AsO_4)_2\cdot8H_2O$ | pale rose | |
| Metalodevite | $Zn(UO_2)_2(AsO_4)_2\cdot10H_2O$ | pale yellow to green | yellowish green (SW,LW) |
| Metanatroautunite[b] | $Na_2(UO_2)_2(PO_4)_2\cdot8H_2O$ | lemon yellow | yellowish green (SW,LW) |
| Metanováčekite | $Mg(UO_2)_2(AsO_4)_2\cdot4\text{-}8H_2O$ | pale yellow to yellow | dull green (SW,LW) |
| Metatorbernite | $Cu(UO_2)_2(PO_4)_2\cdot8H_2O$ | green | |
| Meta-uranocircite I | $Ba(UO_2)_2(PO_4)_2\cdot6\text{-}8H_2O$ | yellow-green | yellow-green (SW,LW) |
| Meta-uranocircite II | $Ba(UO_2)_2(PO_4)_2\cdot6H_2O$ | yellow-green | yellow-green (SW,LW) |
| Meta-uranospinite | $Ca(UO_2)_2(AsO_4)_2\cdot8H_2O$ | lemon yellow to green | bright lemon yellow |
| Metazeunerite | $Cu(UO_2)_2(AsO_4)_2\cdot8H_2O$ | pale green to emerald green | |
| Nováčekite | $Mg(UO_2)_2(AsO_4)_2\cdot12H_2O$ | pale yellow to yellow | yellowish green (SW,LW) |
| Sabugalite | $H_{0.5}Al_{0.5}(UO_2)_2(PO_4)_2\cdot8H_2O$ | yellow | bright lemon yellow (SW,LW) |
| Saleeite | $Mg(UO_2)_2(PO_4)_2\cdot10H_2O$ | yellow | bright yellowish (SW,LW) |
| Sodium uranospinite | $(Na_2,Ca)(UO_2)_2(AsO_4)_2\cdot5H_2O$ | lemon to straw yellow | yellowish green (SW,LW) |
| Torbernite | $Cu(UO_2)_2(PO_4)_2\cdot10\text{-}12H_2O$ | dark emerald green | |
| Trögerite | $(H_3O)_2(UO_2)_2(AsO_4)_2\cdot8H_2O$ | lemon yellow | |
| Uramphite | $(NH_4)_2(UO_2)_2(PO_4)_2\cdot6H_2O$ | pale green to bottle green | yellow-green |
| Uranocircite | $Ba(UO_2)_2(PO_4)_2\cdot12H_2O$ | greenish yellow to yellow | green (SW,LW) |
| Uranospinite | $Ca(UO_2)_2(AsO_4)_2\cdot10H_2O$ | lemon yellow to green | bright lemon yellow |
| Zeunerite | $Cu(UO_2)_2(AsO_4)_2\cdot10\text{-}16H_2O$ | yellowish green to emerald green | |
| **Phosphuranylite Group** | | | |
| Althupite | $AlTh(UO_2)[(UO_2)_3O(OH)(PO_4)_2]_2(OH)_3\cdot15H_2O$ | yellow | |
| Arsenuranylite | $Ca(UO_2)_4(AsO_4)_2(OH)_4\cdot6H_2O$ | orange | |
| Bergenite | $(Ba,Ca)_2(UO_2)_3(PO_4)_2(OH)_4\cdot5H_2O$ | yellow | weak green (SW) |

| Mineral | Formula | Color | UV Fluorescence[a] |
|---|---|---|---|
| Dewindtite[c] | $Pb[H(UO_2)_3O_2(PO_4)_2]_2 \cdot 12H_2O$ | yellow | light green |
| Dumontite | $Pb_2(UO_3)_3O_2(PO_4)_2 \cdot 5H_2O$ | yellow to golden yellow | |
| Françoisite-(Nd) | $Nd(UO_2)_3(PO_4)_2O(OH) \cdot 6H_2O$ | yellow | |
| Hügelite | $Pb_2(UO_2)_3(AsO_4)_2(OH)_4 \cdot 3H_2O$ | brown to orange-yellow | |
| Mundite | $Al(UO_2)_3(PO_4)_2(OH)_3 \cdot 5.5H_2O$ | pale yellow | |
| Phosphuranylite | $KCa(H_3O)_3(UO_2)_7(PO_4)_4$ | deep yellow | |
| Phuralumite | $Al_2(UO_2)_3(PO_4)_2(OH)_6 \cdot 10H_2O$ | lemon yellow | |
| Phurcalite | $Ca_2(UO_2)_3O_2(PO_4)_2 \cdot 7H_2O$ | yellow | yellowish green (SW,LW) |
| Upalite | $Al(UO_2)_3O(OH)(PO_4)_2 \cdot 7H_2O$ | yellow-brown | |
| Vanmeersscheite | $U^{6+}(UO_2)_3(PO_4)_2(OH)_6 \cdot 4H_2O$ | yellow | strong pale green |
| Metavanmeersscheite | $U^{6+}(UO_2)_3(PO_4)_2(OH)_6 \cdot 2H_2O$ | yellow | strong green (SW,LW) |
| Yingjiangite | $K_2Ca(UO_2)_7(PO_4)_4(OH)_6 \cdot 6H_2O$ | yellow to golden yellow | |
| **Walpurgite Group** | | | |
| Hallimondite | $Pb_2(UO_2)(AsO_4)_2$ | yellow | |
| Orthowalpurgite | $(BiO)_4(UO_2)(AsO_4)_2 \cdot 2H_2O$ | pale yellow | |
| Parsonsite | $Pb_2UO_2(PO_4)_2$ | amber to yellow or brownish yellow | |
| Ulrichite | $CaCu(UO_2)(PO_4)_2 \cdot 4H_2O$ | bright apple green to lime green | |
| Walpurgite | $(BiO)_4(UO_2)(AsO_4)_2 \cdot 2H_2O$ | yellow | |
| **Other Phosphates and Arsenates** | | | |
| Arsenuranospathite | $HAl(UO_2)_4(AsO_4) \cdot 40H_2O$ | white to pale yellow | |
| Asselbornite | $(Pb,Ba)(UO_2)_6(BiO)_4(AsO_4)_2(OH)_{12} \cdot 3H_2O$ | brown to yellow | |
| Chistyakovaite | $Al(UO_2)_2[(F,OH)(AsO_4)_2] \cdot 6.5H_2O$ | yellow | |
| Coconinoite | $Fe_2Al_2(UO_2)_2(PO_4)_4(SO_4)(OH)_2 \cdot 20H_2O$ | pale yellow | |
| Furongite | $Al_2(UO_2)(PO_4)_2(OH)_2 \cdot 8H_2O$ | lemon yellow | yellow-green |
| Kamitugaite | $PbAl(UO_2)_5[(P,As)O_4]_2(OH)_9 \cdot 9.5H_2O$ | yellow | |
| Moreauite | $Al_3(UO_2)(PO_4)_3(OH)_2 \cdot 13H_2O$ | yellow-green | |
| Seelite | $Mg(UO_2)_2(AsO_4)_2 \cdot 7H_2O$ | yellow | |
| Triangulite | $Al_3(UO_2)_4(PO_4)_4(OH)_5 \cdot 5H_2O$ | bright yellow | |
| Uranospathite | $HAl(UO_2)_4(PO_4) \cdot 40H_2O$ | yellow to pale yellow-green | yellow-green |
| Przhevalskite | $Pb(UO_2)_2(PO_4)_2 \cdot 4H_2O$ | pale yellow to pale green | |
| Ranunculite | $HAl(UO_2)PO_4(OH)_3 \cdot 4H_2O$ | golden yellow | |
| Threadgoldite | $Al(UO_2)_2(PO_4)2(OH) \cdot 8H_2O$ | yellow | green (SW,LW) |
| Vochtenite | $(Fe^{2+},Mg)Fe^{3+}(UO_2)_4(PO_4)_4(OH) \cdot 12-13H_2O$ | brown | |
| Xiangjiangite | $(Fe^{3+},Al)(UO_2)_4(PO_4)_2(SO_4)_2(OH) \cdot 22H_2O$ | yellow | |
| **Secondary Vanadates** | | | |
| **Carnotite Group** | | | |
| Carnotite | $K_2(UO_2)_2(VO_4)_2 \cdot 3H_2O$ | orange, yellow to greenish yellow | |
| Curienite | $Pb(UO_2)_2V_2O_8 \cdot 5H_2O$ | canary yellow | |
| Francevillite | $Ba(UO_2)_2V_2O_8 \cdot 5H_2O$ | orange-yellow to olive green | |
| Fritzschite | $Mn^{2+}(UO_2)_2[(P,V)O_4]_2 \cdot 10H_2O$ | red-brown to red | |
| Margaritasite | $(Cs,K,H_3O)_2(UO_2)_2(VO_4)_2 \cdot H_2O$ | yellow | |
| Metatyuyamunite | $Ca(UO_2)_2V_2O_8 \cdot 3H_2O$ | yellow | |
| Sengierite | $Cu_2(UO_2)_2(VO_4)_2(OH)_2 \cdot 6H_2O$ | olive green | |
| Strelkinite | $Na_2(UO_2)_2V_2O_8 \cdot 6H_2O$ | golden yellow to yellow-green | dull green |
| Tyuyamunite | $Ca(UO_2)_2V_2O_8 \cdot 5-8H_2O$ | bright yellow to yellow-green | |
| Vanuranylite | $(H_3O,Ba,Ca,K)_{1.6}(UO_2)_2V_2O_8 \cdot 4H_2O$ | yellow | |
| **Other Vanadates** | | | |
| Vanuralite | $Al(UO_2)_2(VO_4)_2(OH) \cdot 11H_2O$ | lemon yellow | |
| Metavanuralite | $Al(UO_2)_2(VO_4)_2(OH) \cdot 8H_2O$ | lemon yellow | |
| Ferghanite | $LiH[(UO_2)(OH)_4(VO_4)]_2 \cdot 2H_2O(?)$ | bright yellow | |
| Rauvite | $Ca(UO_2)_2V_{10}O_{28} \cdot 16H_2O$ | purple to grayish black | |
| Uvanite | $U^{6+}_2V^{5+}_6O_{21} \cdot 15H_2O(?)$ | brown-yellow | |

| Mineral | Formula | Color | UV Fluorescence[a] |
|---|---|---|---|
| **Secondary Silicates** | | | |
| **Uranophane Group** | | | |
| Boltwoodite | $HK(UO_2)SiO_4 \cdot 1.5H_2O$ | yellow to orange | dull green |
| Cuprosklodowskite | $Cu(UO_2)_2Si_2O_7 \cdot 6H_2O$ | grass green to emerald green | |
| Kasolite | $Pb(UO_2)(SiO_4) \cdot H_2O$ | yellow to orange | |
| Sodium boltwoodite | $(H_3O)(Na,K)(UO_2)SiO_4 \cdot H_2O$ | pale yellow | |
| Oursinite | $Co(UO_2)_2Si_2O_7 \cdot 6H_2O$ | pale yellow | |
| Sklodowskite | $Mg(UO_2)_2Si_2O_7 \cdot 6H_2O$ | yellow | |
| Swamboite | $U^{6+}H_6(UO_2)_6(SiO_4)_6 \cdot 30H_2O$ | pale yellow | |
| Uranophane | $Ca(UO_2)_2[SiO_3(OH)]_2 \cdot 5H_2O$ | yellow | very weak yellow-green |
| Uranophane0-ß | $Ca(UO_2)_2[SiO_3(OH)]_2 \cdot 5H_2O$ | yellow to amber | |
| **Weeksite Group** | | | |
| Haiweeite | $Ca(UO_2)_2Si_5O_{12}(OH)_2 \cdot 4.5H_2O$ | pale yellow | dull green |
| Metahaiweeite | $Ca(UO_2)_2(Si_6O_{15}) \cdot H_2O(?)$ | pale yellow to green-yellow | weak green |
| Weeksite | $(K,Na)_2(UO_2)_2(Si_5O_{13}) \cdot 3H_2O$ | yellow | |
| Coutinhoite | $(Th,Ba)_{0.5}(UO_2)_2(Si_5O_{13}) \cdot 1\text{-}3.5H_2O$ | yellow | |
| **Other Silicates** | | | |
| Soddyite | $(UO_2)_2(SiO_4) \cdot 2H_2O$ | canary yellow to amber | |
| Uranosilite | $(UO_2)Si_7O_{15}$ | pale yellow | |
| **Other Minerals Containing Essential U or Th** | | | |
| **Rhabdophane Group** | | | |
| Brockite | $(Ca,Th,Ce)(PO_4) \cdot H_2O$ | yellow to red-brown | |
| Grayite | $(Th,Pb,Ca)PO_4 \cdot H_2O$ | pale yellow to red-brown | |
| Ningyoite | $(U,Ca)_2(PO_4)_2 \cdot 2H_2O$ | brown to brownish green | |
| Tristramite | $(Ca,U^{4+})(PO_4) \cdot 2H_2O$ | pale yellow to yellow-green | |
| **Monazite Group** | | | |
| Cheralite-(Ce) | $(Ce,Ca,Th)(P,Si)O_4$ | green to brown | |
| Monazite-(Ce) | $(Ce,La,Th)(PO_4)$ | brown to greenish white | |
| Monazite-(Sm) | $(Sm,Gd,Ce,Th)(PO_4)$ | pale yellow | |
| **Pyrochlore Group** | | | |
| Betafite | $U^{4+}(Nb,Ti)_2O_6OH$ | brown to black | |
| Uranmicrolite | $(U,Ca,Ce)_2(Ta,Nb)_2(O,OH,F)_7$ | yellow-brown, green-brown to black | |
| Uranopyrochlore | $(U,Ca,Ce)_2(Nb,Ta)_2(O,OH,F)_7$ | amber, brown to black | |
| Bismutopyrochlore | $(Bi,U,Ca,Pb)_{1+x}(Nb,Ta)_2O_6(OH) \cdot nH_2O$ | greenish brown to black | |
| **Samarskite Group** | | | |
| Calciosamarskite | $(Ca,Fe^{3+},U,Y,Th,REE)(Nb,Ta)O_4$ | black | |
| Ishikawaite | $(U,Fe,Y,Ca)(Nb,Ta)O_4$ | black | |
| **Other Minerals** | | | |
| Chevkinite-(Ce) | $(Ce,La,Ca,Na,Th)_4(Fe^{2+},Mg)_2(Ti,Fe^{3+})_3Si_4O_{22}$ | black | |
| Ciprianiite[d] | $Ca_4[(Th,U)(REE)]Alv_2(Si_4B_4O_{22})(OH,F)_2$ | black | |
| Ekanite | $ThCa_2Si_8O_{20}$ | black | |
| Euxenite-(Y) | $(Y,Ca,Ce,U,Th)(Nb,Ta,Ti)_2O_6$ | black | |
| Holfertite | $(UO_2)_{1.75}TiO_4[(H_2O)_3Ca_{0.25}]$ | canary yellow to orange yellow | |
| Iraqite-(La)[d] | $(La,Ce,Th)(Ca,Na)_2(K_{1-x}v_x)Si_8O_{20}$ | yellow-green | |
| Liandratite | $U^{6+}(Nb,Ta)_2O_8$ | yellow to brown | |
| Niobo-aeschynite-(Ce) | $(Ce,Ca,Th)(Nb,Ti)_2(O,OH)_6$ | black | |
| Petschekite | $U^{4+}Fe^{2+}(Nb,Ta)_2O_8$ | black | |
| Polycrase-(Y) | $(Y,Ca,Ce,U,Th)(Ti,Nb,Ta)_2O_6$ | black | |
| Steacyite[d] | $Th(Na,Ca)_2(K_{1-x}v_x)Si_8O_{20}$ | gray to dark brown | |
| Steenstrupine-(Ce) | $Na_{14}Ce_6Mn^{2+}Mn^{3+}Fe^{2+}_2(Zr,Th)$– $(Si_6O_{18})_2(PO_4)_7 \cdot 3H_2O$ | red-brown to black | |
| Tanteuxenite-(Y) | $(Y,Ce,Ca,U)(Ta,Nb,Ti)_2O_6$ | brown to black | |
| Thorbastnaesite | $Th(Ca,Ce)(CO_3)_2F_2 \cdot 3H_2O$ | brown | |
| Thornasite | $(Na,K)ThSi_{11}(O,F,OH)_{25} \cdot 8H_2O$ | pale green to biege | green(SW,LW) |

| Mineral | Formula | Color | UV Fluorescence[a] |
|---|---|---|---|
| Thorosteenstrupine | $Na_{0.5}Ca_{1-3}(Th,REE)_6(Mn,Fe,Al,Ti)_{4-5}-$ | | |
| | $[Si_6(O,OH)_{18}]_2[(Si,P)O_4]_6(OH,F,O)_x \cdot nH_2O$ | dark brown to black | |
| Thorutite | $(Th,U,Ca)Ti_2(O,OH)_6$ | black | |
| Turkestanite[d] | $Th(Ca,Na)_2(K_{1-x}v_x)Si_8O_{20} \cdot nH_2O$ | brown or apple green | |
| Umbozerite | $Na_3Sr_4(Mn,Fe)ThSi_8O_{24}$ | bottle-green to brown | |
| Uranopolycrase | $(U,Y)(Ti,Nb)_2O_6$ | brown-red | |
| Yttrocrasite-(Y) | $(Y,Th,Ca,U)(Ti,Fe^{3+})_2(O,OH)_6$ | black | |

[a]Reported fluorescence can be subjective and to some degree anecdotal; it is rarely studied or reported as precisely as other optical properties. However, UV fluorescence is an important characteristic of quite a few uranium minerals, and many collectors have a general interest in the phenomenon. In the table, SW and LW denote short-wave (254 nm) and long-wave (360 nm) UV respectively, when reported data include this information. In general, many fluorescent uranium minerals will display similar colors in SW and LW but the fluorescence will be more intense under SW illumination than under LW (Frondel 1958). Many of the minerals fluoresce green because this is an intrinsic property of the uranyl ion, which can even be seen when uranyl ions are dissolved in water (Robbins 1994). Copper-containing species such as torbernite, zeunerite, and others are generally not fluorescent because the copper ions quench the fluorescence. Occasionally a specimen will show anomalous fluorescence, either positive or negative. In the case of fluorescent torbernite, the crystals might be zoned, with a thin layer of autunite or other fluorescent member of the group on the surface. Many zippeites fluoresce more weakly than their synthetic counterparts, usually because of admixed soil or clay.

[b]Many references refer to this species as sodium autunite.

[c]Renardite is now regarded as synonymous with dewindtite.

[d]In structural formula "v" denotes a vacancy.

# Index of Mineral Names

*Page numbers in bold indicate photographs*

# Index to Mineral Localities

*Page numbers in bold indicate photographs*